Semantic Properties of Diagrams and Their Cognitive Potentials

Studies in the Theory and Applications of Diagrams

Studies in the Theory and Applications of Diagrams is aimed at the presentation of sustained original work in any area related to research that improves our understanding of the role of diagrams and sketches in communication, cognition, creative thought, learning, and problem-solving.

While diagrammatic representations have been a cornerstone of communication throughout human history, modern media are much more graphical than those of the past. Recent advances in technologies for information presentation—on computers, PDAs and cell phones screens—have immersed us in a reality that is richly pervaded by diagrammatic representations. This has triggered a surge of interest in the study of diagrammatic notations, especially in academic disciplines dealing with cognition, computation and communication, such as human-computer interfaces, information visualization, pure and applied sciences, psychology, education and philosophy. *Studies in the Theory and Applications of Diagrams* aims to be an interdisciplinary series, publishing relevant contributions from any discipline or methodology.

Semantic Properties of Diagrams and Their Cognitive Potentials

Atsushi Shimojima

CSLI PUBLICATIONS
Center for the Study of
Language and Information
Stanford, California

Dedicated to Jon Barwise,
to whom I promised this book.

Center for the Study of Language and Information
Leland Stanford Junior University
Printed in the United States
19 18 17 16 15   1 2 3 4 5

Library of Congress Cataloging-in-Publication Data

Names: Shimojima, Atsushi, 1962- author.

Title: Semantic properties of diagrams and their cognitive potentials / by Atsushi Shimojima, Center for the Study of Language and Information, Leland Stanford Junior University, Stanford, California.

Description: Stanford, California : CSLI Publications, [2015] | Series: CSLI Studies in the Theory and Applications of Diagrams | Includes bibliographical references and index.

Identifiers: LCCN 2015035468 |
ISBN 9781575868493 (pbk. : alk. paper)

Subjects: LCSH: Semantics–Psychological aspects. | Logic diagrams. | Visual communication. | Reasoning (Psychology) | Cognitive science. | Language and logic. | Psycholinguistics.

Classification: LCC P325.5.P78 S55 2015 |
DDC 401/.43–dc23

LC record available at http://lccn.loc.gov/2015035468

CIP

∞ The acid-free paper used in this book meets the minimum requirements of the American National Standard for Information Sciences—Permanence of Paper for Printed Library Materials, ANSI Z39.48-1984.

CSLI was founded in 1983 by researchers from Stanford University, SRI International, and Xerox PARC to further the research and development of integrated theories of language, information, and computation. CSLI headquarters and CSLI Publications are located on the campus of Stanford University.

CSLI Publications reports new developments in the study of language, information, and computation. Please visit our web site at
http://cslipublications.stanford.edu/
for comments on this and other titles, as well as for changes and corrections by the author and publisher.

Contents

Acknowledgments

The theory presented in this book has been around quite a while. It dates back to my 1996 dissertation and less cohesive versions have been presented in various forums. Some parts of the theory have been published as papers.

Here I cannot acknowledge all the people who read or listened to my ideas and gave productive feedbacks in this period. I name just a few individuals: Jon Barwise, who as my dissertation adviser and my first co-author helped me to start the project; Yasuhiro Katagiri, who as my long-term co-author on related themes kept encouraging the development of the theory; Ruth Millikan, who through both her philosophical writings and personal advices had profound effects on the general direction of this project; Dave Barker-Plummer, who offered thorough and penetrating comments on the manuscript during my stay at the Center for the Study of Language and Information, Stanford.

I am also grateful to the two anonymous reviewers of the manuscript for helpful comments, to students who attended my classes and gave fresh insights on the application of the theory, and to my colleagues at Doshisha University who gave me a chance to leave teaching and administrative work to concentrate on writing. This work was also supported by JSPS Grant-in-Aid for Scientific Research (B), Grant Numbers 20500244 and 23300101.

Finally, I give my wholehearted gratitude to my wife Mari, without whose company nothing would have happened in the first place.

March 2015 Atsushi Shimojima

1

Introduction

1.1 General Problem

By a representation system, we mean a set of regularities with which a group of people uses a certain kind of objects to express and extract information about another kind of objects. The first kind of objects are called *representations*, and when they are "graphical" or "diagrammatic," we call the system a *graphical* or *diagrammatic representation system*. We use "graphical" and "diagrammatic" interchangeably throughout this book. Although the scientific concept of "graphical" or "diagrammatic" will not be discussed until later in this book, we start by giving some obvious examples of graphical representation systems.

A group of people use a certain kind of floor maps to express and extract information about floors of buildings, and there are certain regularities with which they express and extract room boundaries, doors, sinks, and other facilities. A different group of people shares a slightly different way of expressing and extracting these items, and in that case, two different sets of regularities hold on the two practices of the two groups, so we consider two different graphical representation systems are involved. The practice may change over time in the same group, or in the process of inheritance by the group's next generation, and in that case, a new set of regularities begins at that point so we again consider two different graphical representation systems are involved.

This way, we can think of various systems of graphical representations. They include the variant systems of floor maps just described, and extend to variants of neighborhood maps, subway route maps, line drawings, feature comparison tables, bar charts, and geometry diagrams. Each system is characterized by a set of regularities on the way representations are used to express and extract information about what they represent.[1]

[1] In some graphical systems, representations appear to be used to express or extract information that is not concerned with any particular objects in the world. We will discuss how such systems can be handled in Chapter 2, where we describe our conceptual framework in greater detail.

Although theoretical linguistics has provided over a century of studies on linguistic representation systems based on English, Japanese, and other natural languages, the history of theoretical studies of graphical representation systems has been very short. Thus, until recently, little had been known about the general properties of graphical representation systems, especially about their cognitive functions that should underlie our everyday experience where some graphical systems are effective for certain cognitive tasks while others are ineffective or even misleading.

Remember what the most recent situation was when you felt compelled to use a diagram to help you think or to better put across your ideas. It might have been when you tried to come up with a geometry proof as a high school student, or when you explained a geometry proof to your students; it might been when you designed your garden, or when you explained your design idea to your spouse; it might have been when you tried to analyze the annual sales of your company, or when you explained your analysis to your colleagues; it might been when you tried to find your way to a shopping center, or when you were asked how to get there. In many of these situations, graphical representations had a dramatic effect on what you were trying to do. Yet there had been no branch of science that explained why they had that effect. You might have even experienced that a slight change in graphical notations made a significant difference. Science had offered no explanation for this difference, either. While graphical representations are produced and used everyday, human beings had no reference as to when graphics facilitated, hindered, or misguided our thinking and understanding.

1.2 Origin of This Book

The situation started changing, though gradually, when Jaque Bertin started comparing different systems of graphical representations to obtain a generalization on what made a particular system suitable for expressing a certain set of data (Bertin 1967/1973; 1977/1981). Edward Tufte (1983; 1990) continued what Bertin had started, elaborating Bertin's insights and adding his own generalizations. Both Bertin's and Tufte's work reflected what a modern terminology would dub the "information design" concern to find theoretically sound principles for designing effective information graphics. This concern was shared by such studies as those by Cleveland (1994) and Kosslyn (1994), although they were based on a psychological point of view.

Somewhat independently of this line of study, a mixture of cognitive scientists, AI researchers, philosophers, and logicians started intensively studying diagrammatic representation systems. Although the earliest contributions date back to the 1950s (e.g., Gelernter 1959), the movement significantly accelerated around 1990, when the AI researchers' interests in utilizing diagrams for automated reasoning (e.g., Lindsay 1988/1995; Novak 1988/1995) productively converged with the cognitive psychologists' interests in investigating human cognitive processes utilizing diagrams (e.g., Larkin and Simon 1987/1995; Hegarty 1992/1995) and the logicians' interests in formalizing the deductive inferences enabled by diagrams (e.g., Barwise and Etchemendy 1991; Shin 1994). Thus, the researchers involved in this movement were ostensively interested in the functional roles of diagrammatic notations in human cognitive processes.

As studies accumulated along these lines of research, it became increasingly clear that several themes recurred again and again in their discussions. Diagrams prompt, enable, or mandate some characteristic cognitive processes to perform certain cognitive tasks, and apparently, these characteristic patterns, which we will call *cognitive potentials of diagrams*, are common to many different graphical systems. They include positive potentials such as diagrams' letting us skip some deductive skips in inferential tasks, as well as negative potentials such as diagrams' mandating the expression of overly specific information. These cognitive potentials, recurrently discussed in the literature, constitute the main themes of this book.

In the following, we will give a brief preview of each of these themes. We will illustrate them with a simple system of position diagrams to show the bare bones of each phenomenon, although the actual discussions in the literature have rarely employed such simplistic examples.

1.3 Main Theme 1: Free Rides

Suppose we express the following information about baseball players both in a position diagram and in an English text:

(θ_1) Mantle is better than Mays.

Figure 1-a is the expression in a position diagram and Figure 1-b is the typical expression in an English text. In this system of position diagrams, names of people are supposed to appear in a horizontal arrangement without overlaps or repetitions. These rules are called *syntactic* stipulations in the sense that they are concerned with the *forms* in which symbols should be combined in a representation. In contrast, *semantic* conventions are concerned with the *meanings* of these forms, specifying what configurations of symbols indicate what information about the represented situation. The basic semantic convention for this system of position diagrams says:

(1) If the name of player X appears to the left of the name of player Y, it indicates that X is better than Y.

Thus, since Mantle's name appears to the left of Mays' name in Figure 1-a, the position diagram expresses information θ_1. The English sentence in Figure

1-b expresses information θ_1 according to its usual semantic conventions.

(a) Mantle Mays

(b) Mantle is better than Mays.

FIGURE 1 The expression of information θ_1 in a position diagram (a) and in an English text (b).

The two representations in Figure 1 look similar with both consisting of a linear sequence of symbols. However, this is so because only a single piece of information θ_1 is expressed. Crucial differences between the two representation systems reveal themselves in the ways they handle multiple pieces of information.

Suppose we change the position diagram and the English sentence to express one more piece of information θ_2:

(θ_2) Mays is better than Moskowitz.

According to semantic convention (1) for the position diagrams specified above, this requires Mays' name to appear to the left of Moskowitz's name. Therefore, we place Moskowitz's name to the right of Mays' name. As for the English text, we need to add another sentence expressing θ_2. Thus, we will have the modified position diagram and English text shown in Figures 2-a and 2-b.

(a) Mantle Mays Moskowitz

(b) Mantle is better than Mays. Mays is better than Moskowitz.

FIGURE 2 The expression of information θ_1 and θ_2 in a position diagram (a) and in an English text (b).

Now, let us see what happened to the names in the position diagrams. As Moskowitz's name is placed to the right of Mays' name while Mantle's name is placed to the left of Mays' name, Mantle's name necessarily comes to the left of Moskowitz's name. According to semantic rule (1) for position diagrams, this indicates the following information:

(θ_3) Mantle is better than Moskowitz.

Note that θ_3 is a different piece of information from information θ_1 and θ_2 that we are supposed to express in the position diagram. Yet, it is automatically expressed in the position diagram upon the expression of θ_1 and θ_2. Furthermore, θ_3 is a *consequence* of information θ_1 and θ_2: whenever θ_1 and θ_2 hold of a situation, θ_3 must hold of that situation. Thus, expressing a certain set of information in a position diagram results in expressing other, consequential information.

In contrast, the English text does nothing beyond what it is supposed to do. We expressed two pieces of information θ_1 and θ_2 in English, and the resulting text in Figure 2-b only express these pieces of information and nothing more. It does not express θ_3, in particular. Of course, once one reads off information θ_1 and θ_2 from the text, it will be just a short step to reach conclusion θ_3. However, this is based on an *inference* from the pieces of information expressed in the text, and must be distinguished from the case where one reads off information θ_3 directly from the text. One could do the latter only if the text contained another English sentence that was dedicated to the expression of θ_3. The text has no such sentence, and does not let one read off information θ_3 as the position diagram does.

Since a position diagram that expresses information θ_1 and θ_2 automatically expresses information θ_3, somebody expressing information θ_1 and θ_2 in a position diagram does not have to deduce θ_3 in his or her head. As the information is expressed there, one can simply read it off. In other words, the system of position diagrams enables one to substitute an inferential task with the task of comprehending θ_3 from a diagram. The option of such a task-substitution is not available for someone expressing information θ_1 and θ_2 in an English text.

We can describe the general potential of diagrams illustrated by this example in the following way:

> **Potential for Free Ride in Inference** Expressing a set of information in diagrams can result in the expression of other, consequential information. This enables us to skip certain mental deductive steps and to substitute them with the task of comprehending the consequences from the diagrams.

1.4 Main Theme 2: Over-Specificity

Assuming the same system of position diagrams, suppose we are now to express the following slightly different pairs of information both in a position diagram and in an English text:

(θ_1) Mantle is better than Mays.

(θ_3) Mantle is better than Moskowitz.

According to semantic rule (1) for position diagrams, expressing θ_1 requires us to place Mantle's name to the left of Mays' name. This is easy. A problem arises when we try to add information θ_3. It requires Mantle's name to appear to the left of Moskowitz's name, but there are two different ways to do so: one is to place Moskowitz's name between Mantle's and Mays' (Figure 3-a), and the other is to place Moskowitz's to the right of Mays' name (Figure 3-b).

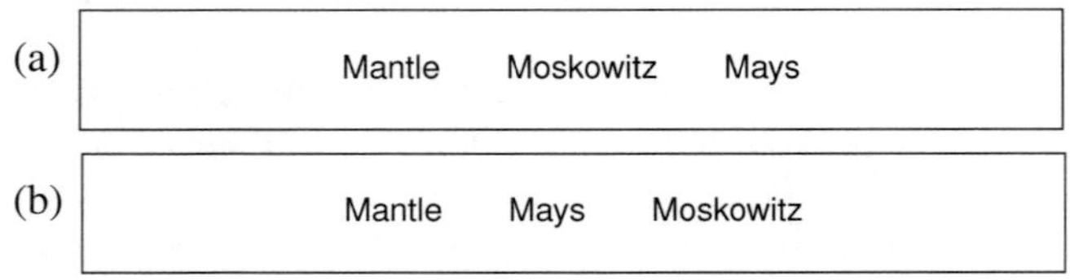

FIGURE 3 Alternative ways of expressing information θ_1 and θ_3 in position diagrams.

The problem is that we generally have no grounds for choosing either alternative. Taking the first alternative forces us to express information θ_6 and taking the second alternative forces us to express information θ_2, but we have no grounds to believe either information to be true:

(θ_6) Moskowitz is better than Mays.

(θ_2) Mays is better than Moskowitz.

In other words, the system of position diagrams does not allow us to express the pair of information θ_1 and θ_3 exclusively, without expressing another unwarranted piece of information. Thus, we are forced to either give up expressing this pair of information or to express it along with a non-warranted piece of information.

In contrast, no such problem occurs in the case of English texts. English allows us to express the pair of information θ_1 and θ_3 exclusively, without expressing any other information. We simply produce a sentence expressing each of θ_1 and θ_3, as in Figure 4, and neither unwarranted information θ_6 nor θ_2 gets expressed. As was previously explained, English only expresses what it is supposed to express, so to speak, and this time, it serves to avoid a problem the system of position diagrams incurs.

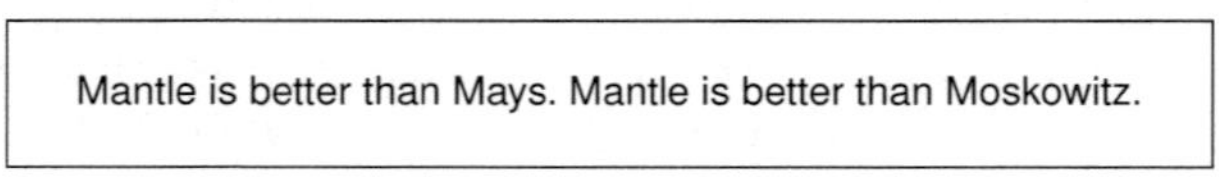

FIGURE 4 The expression of information θ_1 and θ_3 in an English text.

Potential of Over-Specificity Expressing a set of information in diagrams can mandate the selective expression of other, often non-consequential pieces of information. This forces us to either refrain from expressing that set of information in diagrams or tolerate the presence of the additional pieces of unwarranted information in our diagrams.

1.5 Main Theme 3: Consistency Check

Suppose this time we are to express the following three pieces of information in a position diagram and an English text:

(θ_1) Mantle is better than Mays.

(θ_2) Mays is better than Moskowitz.

(θ_{10}) Moskowitz is better than Mantle.

It is easy to produce an English text that expresses these pieces of information. Simply keep adding a sentence expressing each of θ_1, θ_2, and θ_{10} and we obtain the English text that expresses them (Figure 5).

> Mantle is better than Mays. Mays is better than Moskowitz. Moskowitz is better than Mantle.

FIGURE 5 The expression of set of information $\{\theta_1, \theta_2, \theta_{10}\}$ in an English text.

In contrast, it is impossible to produce a position diagram that expresses all of θ_1, θ_2, and θ_{10}. Suppose we start by expressing θ_1 to obtain a position diagram where Mantle's name appears to the left of Mays' name. Now, expressing θ_2 requires Mays' name to appear to the left of Moskowitz's name in this diagram, whereas expressing θ_{10} requires Moskowitz's name to appear to the left of Mantle's name. Since the syntactic stipulations for position diagrams require all names to appear in a horizontal arrangement without repetition, such an arrangement of names is impossible (see Figure 6). This holds in whatever order we express θ_1, θ_2, and θ_{10}.

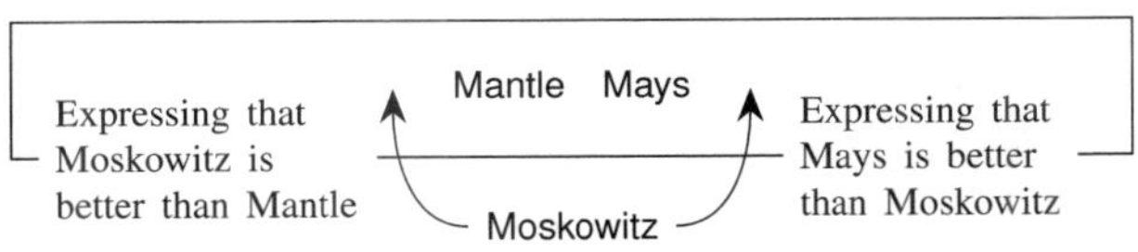

FIGURE 6 Impossibility of expressing set of information $\{\theta_1, \theta_2, \theta_{10}\}$ in a position diagram.

Note that this impossibility corresponds to the fact that set of information $\{\theta_1, \theta_2, \theta_{10}\}$ is inconsistent—i.e., θ_1, θ_2, and θ_{10} cannot hold together of a single situation. If Mantle is better than Mays in some situation, Moskowitz cannot be both worse than Mays and better than Mantle in that situation. There are other inconsistent sets of information that the system of position diagrams cannot express, such as the set of two pieces of information consisting of the information that Mantle is better than Mays and the information that Mays is better than Mantle, and the set of one piece of information (unit set) consisting of the information that Mays is better than Mays.

This illustrates the fact that position diagrams cannot express a certain range of inconsistent sets of information. This limitation, however, implies a distinctive cognitive advantage. Precisely because position diagrams cannot express a certain range of inconsistencies, one can always be sure that a set of information does not fall in this range of inconsistencies as long as one can draw a position diagram that expresses that set of information. In other words, drawing a position diagram serves as some basic consistency check of the set of information it expresses. It therefore substitutes for the inferential process necessary to evaluate its consistency.

Such an alternative method of checking consistency is not available for someone using English texts. This is illustrated by the English text in Figure 6, which expresses the inconsistent set θ_1, θ_2, and θ_{10} with no uncertainty. Producing an English text that expresses a set of information is generally no assurance, to whatever extent, that that set is consistent.

> **Potential for Consistency Check** It is not possible to express a certain range of inconsistent sets of information in diagrams. Thus, drawing a diagram can serve as the basic consistency check of a set of information that the diagram expresses, substituting for the mental process of evaluating the consistency of the set in question.

1.6 Main Theme 4: Derivative Meaning

Figure 7-a shows a position diagram expressing the four pieces of information below, and Figure 7-b shows an English text expressing the same set of information.

(θ_1) Mantle is better than Mays.

(θ_2) Mays is better than Moskowitz.

(θ_{13}) Musial is better than Maris.

(θ_{14}) Moskowitz is better than Musial.

Assuming we are only interested in the players mentioned in these pieces of information, how many players are better than Moskowitz? For someone using the position diagram in Figure 7-a, perhaps the easiest way of answering this question is to count the number of names to the left of Moskowitz's name. Well, there are two names, and this means that there are two people better than Moskowitz. Thus, the answer can be directly read off from the position diagram.

In the case of the English text in Figure 7-b, however, counting the number

(a) Mantle Mays Moskowitz Musial Maris

(b) Mantle is better than Mays. Mays is better than Moskowitz
Musial is better than Maris. Moskowitz is better than Musial.

FIGURE 7 The expression of information θ_1, θ_2, θ_{13}, and θ_{14} in a position diagram (a) and in an English text (b).

of names does not help us answer the question. It is simply not a function of an English text to have the number of names mean anything about the situation it describes. Therefore, we usually go through a rather lengthy process of (i) interpreting each of the three sentences in the text, (ii) integrating the individual pieces of information thus obtained, and (iii) drawing a conclusion appropriate to the question.

Note that the direct reading process illustrated above for the position diagram in Figure 7-a depends on a meaning relation that *goes beyond* basic semantic convention (1) we have assumed for position diagrams. The semantic convention we assumed was:

(1) If the name of player X appears to the left of the name of player Y, it indicates that X is better than Y

The instance of meaning relation underlying the direct reading process was:

(2) If two names are to the left of the name of Y, it indicates that two people are better than Y

It turned out that although (2) is different from (1), it can be *derived* from (1), given the various constraints holding on position diagrams and the situations they represent. The following are other instances of the derivative meaning relation holding in this system of position diagrams:

(3) If more names are to the left of the name of X than to the left of the name of Y, it means that more people are better than X than people who are better than Y

(4) If the name of X is at the leftmost position, it means that X is the best player

Strictly speaking, the following is also different from the basic meaning relation (1) we have assumed:

(5) If the name of X is to the right of the name of Y, it means that X is worse than Y

Using the meaning relation instantiated in (3)–(5), we can extract the following pieces of information from Figure 7-a:

(θ_{15}) More people are better than Musial than people who are better than Mays.

(θ_{16}) Mantle is the best player.

(θ_{17}) Moskowitz is worse than Mays.

The derivative meaning relation is again not available when we read English texts. With an additional meaning relation of this kind, a graphical system with a relatively simple set of basic semantic conventions can have extended expressive coverage.

> **Potential for Derivative Meaning** Graphical systems can support meaning relations that go beyond their basic semantic conventions. This enables us to read off a richer set of information directly from graphics, expanding the expressive coverage of a system with a relatively simple set of basic semantic conventions.

1.7 Organization of This Book

Thus, we have illustrated a total of four cognitive potentials of graphical representation systems. It turned out that each of these potentials is rooted in a more general, *semantic property* of the relevant graphical system, viz., a characteristic way in which properties of diagrams correspond to properties of situations they represent. Furthermore, these semantic properties are *logically characterizable*, in the sense that they can be characterized with reference to various constraints that hold on the key components of a graphical system.

Consequently, we will be engaged in two kinds of activities in this book: (1) pinning down the cognitive potentials of diagrams that have been discussed recurrently but rather sporadically in the previous research, and (2) explaining how a graphical system has each of these potentials by specifying the semantic property underlying it. Each of Chapters 2–5 runs this procedure on each of the four cognitive potentials. Altogether, the result will be a pretty comprehensive list of cognitive potentials of graphical systems, with their underlying semantic mechanisms having been clarified.

It turned out that one of these semantic mechanisms, which we call "consequence tracking," underlies several other cognitive potentials of diagrams than the ones covered in Chapters 2–5. Chapter 6 explores these other cognitive potentials. Although our account in that chapter is still in the stage of an extended hypothesis, it provides useful perspectives from which we may extend the analyses in this book to obtain a still more comprehensive account of the cognitive potentials of diagrams. Our long-term goal is to promote our knowledge of when, how, and why graphical systems facilitate, hinder, or misguide our thinking and understanding. If successful, this book should provide a good map in which different graphical systems can be placed. The semantic properties will serve as the basic coordinates for the map, attributing different cognitive potentials to graphical systems according to where they are placed in the map.

The final chapter reviews the four semantic properties identified in this book, and points out a common generic structure shared by them. The existence of this commonality among the key properties of diagrams that account for many of their cognitive potentials strongly suggests that there is a substance to the things that we study, i.e., the things commonly called "diagrams," that puts them together as a coherent class of objects. On these grounds, we will suggest that diagrams make up a "natural kind" over which we can project a science, rather than a conglomerate of unrelated objects that shares little more than the common denomination.

Since the logical characterization of semantic properties of graphical systems is an essential part of this book's program, a set of logical and set-theoretical concepts (such as *constraints* and *projection*) will be introduced, with basic knowledge of set theory assumed. However, the description in the main body of the book is informal, with most of the formal details kept in the background. Whenever a technical concept needs to be introduced, it will be accompanied by clear examples.

An important feature of this book is that its main content is rooted in the accumulation of research in the information-design tradition and the cognitive-scientific tradition outlined above. Each of our main analyses starts with an examination of examples and suggestions given in the previous research, as well as the identification of a common theme, i.e., a common cognitive potential, indicated by them. As such, a large part of our theory should be taken as a rational reconstruction of the previous work, rather than someone's exclusive creation. This fact lends significant credibility to the importance of the themes that have been chosen for discussion in this book. For this reason, fairly detailed documentation on previous research is included in each relevant chapter.

1.8 The Reach of This Book

We have defined a "cognitive potential of a diagram" as a characteristic process that is prompted, enabled, or mandated by the diagram to perform a certain cognitive task. One could describe this process at different levels of specificity. The level of specificity achieved in this book is rather low. "Free rides," "over-specificity," "consistency check," and "meaning derivation" are all designations of cognitive potentials of diagrams, but our characterizations of the designated processes are far from exact specifications, with no attempts to identify the sub-processes and cognitive resources involved (except for the case of "envisaging" discussed in Chapter 6, whose process model is specified and experimentally tested in Shimojima and Katagiri (2013)). This is mainly because the body of diagrams studies from which we have extracted our main themes are still in its infancy as a scientific field.

Our expectation is that by mapping out major cognitive potentials of diagrams even at this low level of specificity, we may prompt more detailed and focused investigations on each cognitive potential. Informed by findings accumulated in cognitive psychology on human cognitive resources, such investigations would fill the gaps in our low-level specifications to provide experimentally testable process models.

Thus, the cognitive potentials characterized in this book should be considered broad hypotheses, backed up by semantic analyses of varying diagrammatic systems. We expect that they work as intermediaries between logico-semantic studies of diagrammatic systems and specialized cognitive research on the hypothesized processes. The entire book is an extended demonstration that these different traditions of research can be fruitfully combined to achieve a fuller understanding of the cognitive functions of diagrams.

2

Potential for Free Ride in Inference

> Expressing a set of information in diagrams can result in the expression of other, consequential information. This enables us to skip certain mental deductive steps and to substitute them with the task of comprehending the consequences from the diagrams.

In this chapter, we will identify and analyze one central cognitive potential of diagrams that has been repeatedly discussed in previous studies. It is the potential for a free ride in inference—quoted above, and briefly illustrated in Chapter 1. In Section 2.1, we will visit various places in the literature where this potential is described or suggested, and show that they are all concerned with one common phenomenon. A particular theoretical framework will be used for analysis in this and subsequent chapters, so we will devote Section 2.2 to articulating the basic assumptions that make up this framework. After this preparation, Section 2.3 will revisit the individual examples discussed in Section 2.1 and identify the semantic property of a graphical system that accounts for its free ride potentials. Section 2.4 will clarify some key points of our analysis in the form of questions and responses.

2.1 The Phenomenon

2.1.1 Hohauser (1982): Memory Maps

In his book on architectural modeling, Hohauser (1982) cites one paradigmatic example that exhibits what we have called a free ride in inference (p. 177). It is a story about an elderly man, named Harry Lieberman, who constructed a "memory map" of his home village with the help of an artist. The following is our own reconstruction of the story (rather than a historically accurate report) designed to highlight the point that we are making.

Harry is asked to describe the geographical features of the village in which he grew up, as accurately as possible, by memory. After some failed attempts to recollect the geographical features of the village with no external tools, he decides to draw an approximate map of his hometown. On the basis of

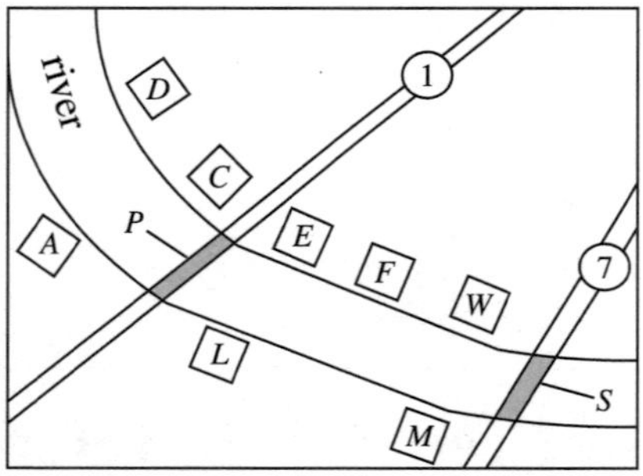

FIGURE 8 Harry's memory map, before information θ_1 was expressed.

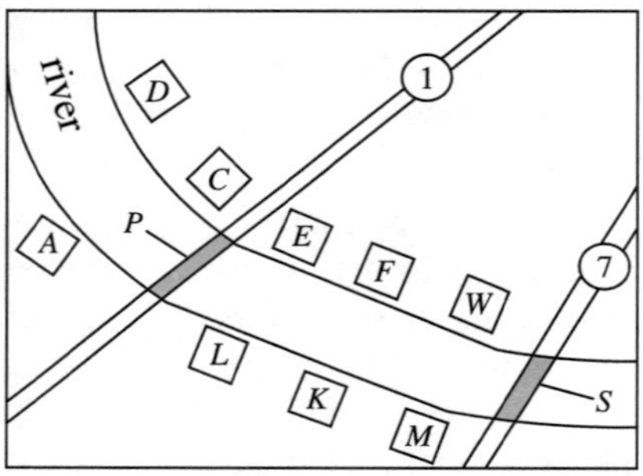

FIGURE 9 Harry's memory map, after information θ_1 was expressed.

fragments of his memory, he draws lines and curves on a sheet of paper to represent the streets, pathways, rivers, and such. He then uses wood blocks to represent the buildings that he remembers having existed, and places them on his map, to represent the approximate locations of those buildings. (He keeps revising and supplementing the map, and eventually obtains a map that represents his hometown to the best of his memory.)

At the beginning of this procedure, Harry remembers the locations of a river, two roads, and several houses, and constructs the tentative map in Figure 8. Then he recollects one more piece of information about his home village, that is:

(θ_1) House K was halfway between houses L and M.

To express this new fragment of memory in his map, Harry puts a wood block standing for house K between the wood blocks standing for houses L and M. Figure 9 shows the resulting map.

As the result of this simple operation, Harry's map now presents many pieces of new information, *other than* θ_1, that were absent from the initial map. Among them are:

(θ_2) House K is across the river from house F.

(θ_3) House K is closer to road 1 than house M is.

(θ_4) House M is closer to bridge S than house K is.

(θ_5) House K and house A have road 1 in between.

To get a sense of the utility of the system of Harry's memory map, imagine how many deduction steps would be needed if he tried to obtain the same results with pure thought on the basis of the principles of geometry. By operating on his map in the way described above, Harry has skipped all these complications of computation, and obtained information θ_2, θ_3, θ_4, and θ_5 almost "for free." The operation is extremely efficient, for the purpose of updating the information content of his map toward the solution of the problem.

2.1.2 Barwise & Etchemendy (1991): Hyperproof Diagrams

Barwise and Etchemendy (1991) are concerned with the same kind of efficacy when they make the following general remark on the utility of diagrams in reasoning:

> Diagrams are physical situations. They must be, since we can see them. As such, they obey their own set of constraints. In our example from *Hyperproof*, when we represent tetrahedron *a* as large, a host of other facts are thereby supported by the diagram. By choosing a representational scheme appropriately, so that the constraints on the diagrams have a good match with the constraints on the described situation, the diagram can generate a lot of information that the user never need infer. Rather, the user can simply read off facts from the diagram as needed. (Barwise and Etchemendy 1991: p. 22.)

Although this passage already contains the idea of matching of constraint, which is of central importance to our analysis, let us now focus on the phenomenon that they are pointing to. Here is the example that they mention as "our example from *Hyperproof*." We are given the picture in Figure 10 that depicts the sizes, shapes, and locations of different blocks on a chess board[2].

We interpret this diagram as we do for ordinary pictures. Namely, the large cube icon on the leftmost column of the grid picture represents an (actual) large cube on the corresponding position on the chess board, the small dodecahedron icon labeled "*c*" represents a small dodecahedron, named "*c*," on the corresponding position on the chess board. Note that the cylinder icon with a triangle symbol represents the tetrahedron *a* whose size is unknown. So, this picture has an element of indeterminacy, unlike most realistic drawings. Now,

[2]Barwise and Etchemendy (1994) have developed an interactive computer program, called "Hyperproof," for teaching basic principles of analytical reasoning. Figure 10 is a typical picture that shows up in a proof that a student constructs with the program.

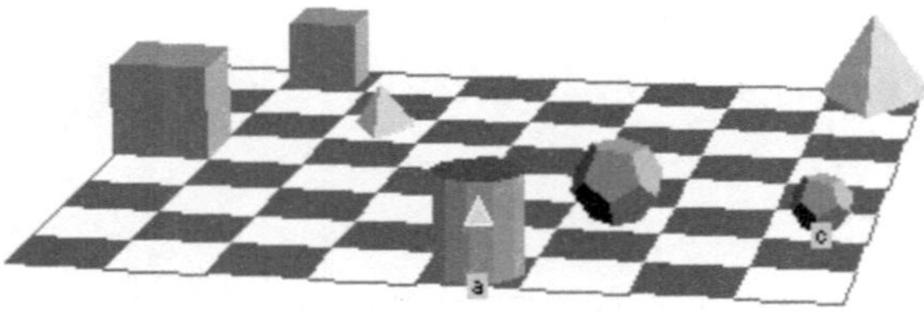

FIGURE 10 Hyperproof diagram, before information θ_6 was expressed.

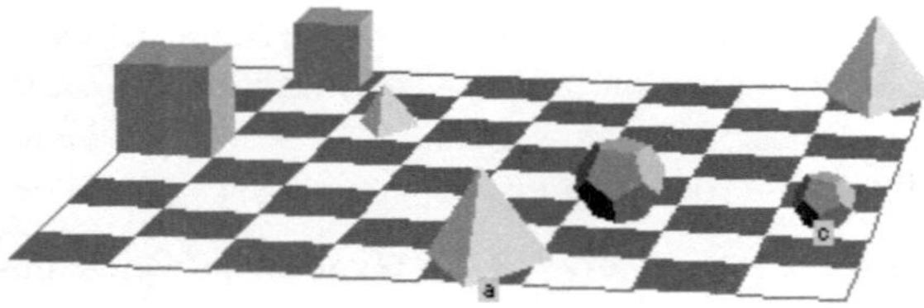

FIGURE 11 Hyperproof diagram, after information θ_6 was expressed.

we change the cylinder icon in this diagram to a large tetrahedron icon, to add the following information:

(θ_6) *a* is large.

As a result, we obtain Figure 11.[3]

In addition to information θ_6, our picture now presents the following information (among others) that was absent from the first picture:

(θ_7) *a* is larger than *c*.

(θ_8) *a* is of the same size as the cube on the leftmost column.

(θ_9) Every object in front of *c* is a large tetrahedron.

(θ_{10}) There are exactly two large tetrahedra.

(θ_{11}) There are no medium sized tetrahedra.

In this example, a relatively small operation of adding a single piece of information θ_6 makes the resulting picture "generate a lot of information," as Barwise and Etchemendy put it. This generation of information saves, in turn, a substantial number of inferences on our part—we "never need infer" information θ_7, θ_8, θ_9, θ_{10}, and θ_{11}, but "simply read [them] off" from the resulting diagram.

[3]This operation would take a substantial effort if the diagram were drawn on a piece of paper, rather than on a computer display where we can use the diagram editor of *Hyperproof*. But this is not relevant to the point that we are making.

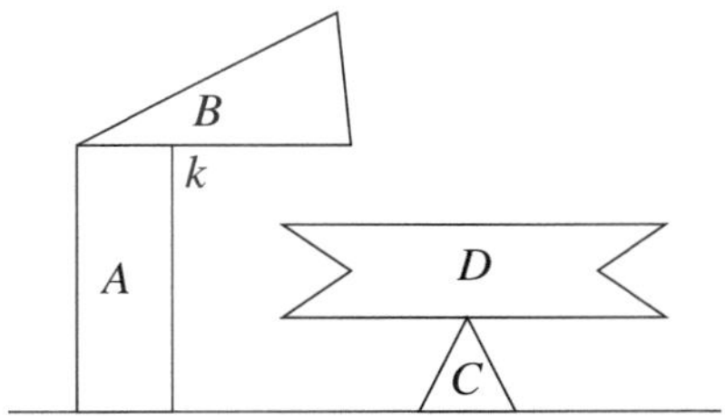

FIGURE 12 Block drawing depicting the state of blocks before *B*'s rotation.

2.1.3 Funt (1980): Block Drawings

Funt (1980/1995) points to the same general phenomenon when he discusses "experimental feedbacks from diagrams." Funt presents a program, called WHISPER, that can observe and modify diagrams to solve certain problems about the motion of collapsing objects[4]. He characterizes WHISPER in the following way:

> Usually feedback is thought of in terms of a robot immersed in a real world environment. In WHISPER's case, however, the feedback is from a situation analogous to that in the real world—the diagram and diagram transformations—rather than from observation of actual falling objects.... Using this feedback WHISPER is able to find when and where discontinuous changes in an object's motion occur without being forced to use sophisticated, "number-crunching" algorithms for touch tests for arbitrary shapes. (Funt 1980/1995: p. 51.)

One of the "experiments" that Funt refers to is the following. WHISPER is given a diagram of a structure built of four blocks, *A*, *B*, *C*, and *D*, that is about to collapse (Figure 12). The blocks are assumed to have uniform density and thickness, and the problem is to predict the sequence of events occurring during the presumed collapse.

As the qualitative solution of the problem, WHISPER outputs a set of diagrams that depict a sequence of events during the collapse. Observing the first diagram with its "retina," WHISPER determines that *B* is in the dominant instability, and predicts a clockwise rotation of *B* on pivot point *k*. WHISPER figures that this rotation of *B* terminates when:

(θ_{12}) The bottom surface of *B* inclines clockwise at 28°.

WHISPER changes the initial diagram so that it may depict this prediction (Figure 13).

The resulting diagram expresses the following information:

(θ_{13}) There is a gap between the lowest edge of *B* and the surface of *D*.

[4]Funt describes, in some detail, how WHISPER controls a machine to "observe" and draw diagrams, but we are not concerned with these issues here.

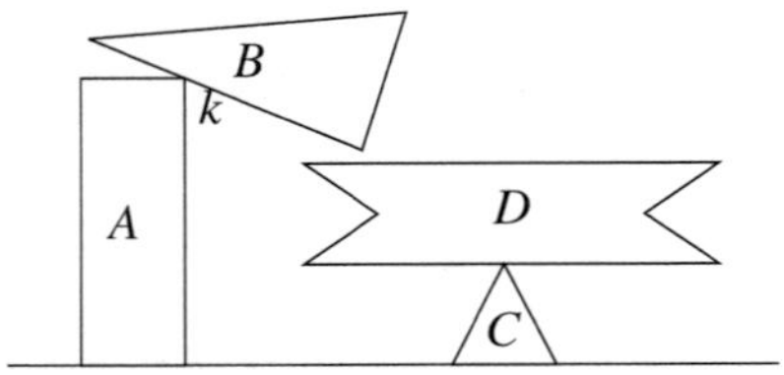

FIGURE 13 Block drawing expressing the terminal state of *B*'s predicted rotation.

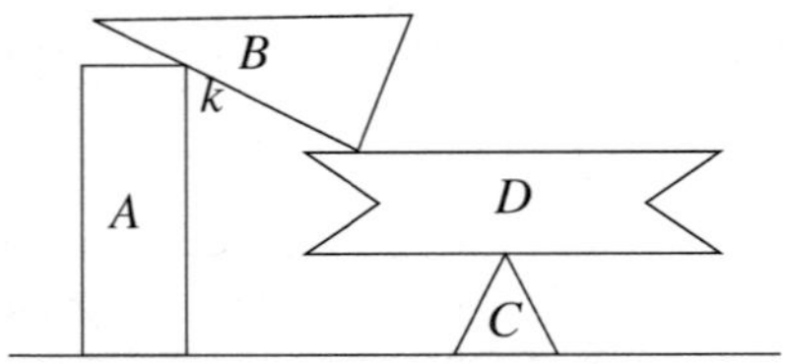

FIGURE 14 Block drawing expressing the result of WHISPER's corrected calculation about *B*'s rotation.

On the basis of this information, WHISPER corrects its computation of *B*'s rotation, and outputs a "correct" diagram that depicts the termination condition of *B*'s rotation (Figure 14).

Repeating this trial and error, WHISPER produces two more snapshots of the presumed collapse by the end of its problem-solving procedure (Figure 15).

Note that θ_{12} and θ_{13} are distinct pieces of information. Information θ_{13} is something that gets expressed in the diagram *as a result of* changing the original diagram to reflect information θ_{12}. In Funt's terminology, θ_{13} is the "experimental feedback" of changing the diagram on the basis of information θ_{12}. Relying on this feedback, WHISPER modifies its calculation of the initial rotation of *B*. According to Funt, this experiment on the diagram spares WHISPER the "sophisticated, 'number-crunching' algorithms" that would be necessary to derive θ_{13} from θ_{12} in the given circumstance. WHISPER repeats similar experiments in the process of producing snapshots in Figure 15.

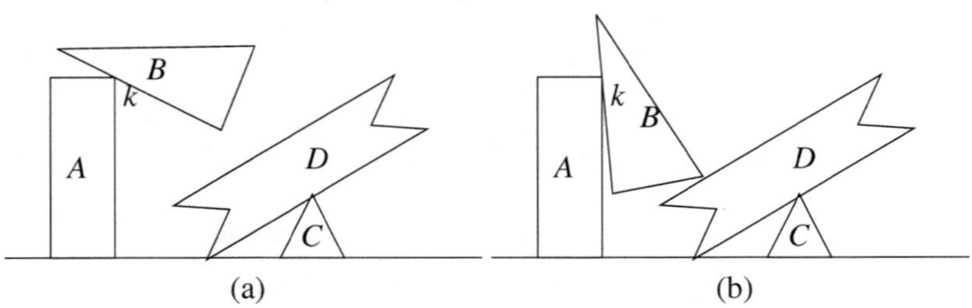

FIGURE 15 Block drawing depicting instantaneous states of the presumed collapse of blocks.

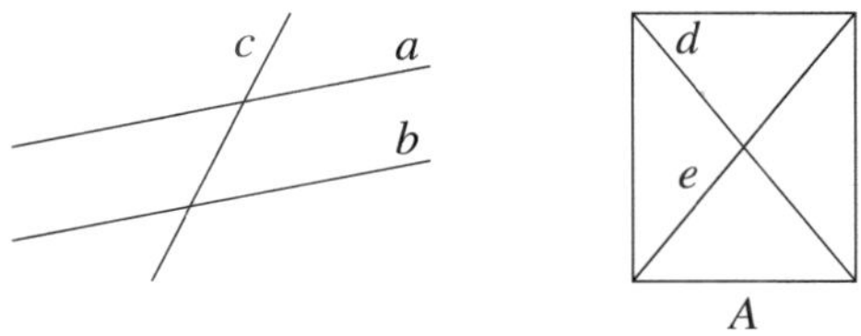

(a) (b)

FIGURE 16 Geometry diagrams supporting free rides .

Thus, WHISPER's experiments are again cases in which an action that expresses a piece of information in a representation generates an independent piece of information in the same representation, which in turn saves a significant amount of inferential work on the part of the user. Funt's example is different from Hohauser's and Barwise and Etchemendy's in that some of the snapshots produced by WHISPER depict different stages of the collapse of the blocks, rather than one and the same state of these blocks. However, the free ride that WHISPER obtains in drawing an *individual* diagram is about a single situation of the blocks. For example, information θ_{12} and θ_{13} are both about the same situation of the blocks, although the situation is an instantaneous one.

2.1.4 Larkin & Simon (1987): Geometry Diagrams

We also find a reference to the same general phenomenon in Larkin and Simon (1987/1995). They note two facts: (a) drawing a pair of parallel lines cut by a transversal on paper generates the information that there are eight angles, four exterior and four interior, defined by the corresponding parallel lines and transversal (Figure 16-a), and (b) drawing a rectangle and its two diagonals on paper generates the information that there is a point of intersection of the corresponding diagonals (Figure 16-b).

Larkin and Simon write:

> The process of drawing the diagram makes these new inferences which are then displayed explicitly in the diagram itself.... Of course, the same information can also be inferred from the sentential representation, but these latter inference processes may require substantial computation, and the cost of this computation must be included in any assessment of the relative efficiency of the two representations. (Larkin and Simon 1987/1995: p. 74.)

Lindsay (1988/1995) also takes up the case of geometry diagrams, and discusses their ability to "permit the drawing of inferences without explicit use of rules of deduction" (p. 112). He calls a representation system with this capacity "non-deductive," and says that a nondeductive system "requires no separate computational inference-making stage: the operation of the construc-

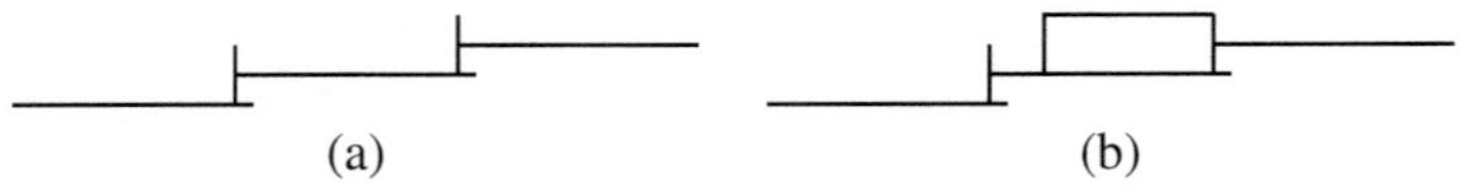

FIGURE 17 Schematic reproduction of (a) Quist's first sketch and (b) Quist's modified sketch.

tion process entails the 'making' of the inferences" (p. 115).

Mumma (2010) offers another clear example of free ride made with geometry diagrams (p. 279). Just draw a standard geometry diagram to express the information that point *B* lies between points *A* and *C* and that point *C* lies between points *B* and *D*. Your diagram inevitably expresses the additional, consequential information that *C* lies between *A* and *D*. as simple as it is, this example is rather difficult to grasp without actually drawing the diagram. It is therefore a good demonstration of the importance of physical drawing, and particularly of a free ride, in a geometrical inference.

2.1.5 Schön (1983): Design Sketches

In his classical protocol-analysis of the design practice of architects, Schön (1983) extensively discusses an example where an architect, Quist, was helping an architectural student, Petra, with the design of elementary-school buildings on a land with slopes. Quist's use of design sketches in this example produced what we have called free rides, and they helped him and Petra to see the consequences of their design ideas and to discover new ideas.

Schön describes this example in the following way:

> Each move is a local experiment which contributes to the global experiment of reframing the problem. Some moves are resisted (the shape cannot be made to fit the contours), while others generate new phenomena. As Quist reflects on the unexpected consequences and implications of his moves, he listens to the situation's back talk, forming new appreciations which guide his further moves. (Schön 1983: p. 94.)

One of the unexpected consequences mentioned in the above quote was generated in the following way.[5] After listening to Petra's initial idea of the structure of class-room buildings, Quist took the lead of thinking of how to realize her idea on the land with a problematic slope. Quist drew a sketch that depicts, as in Figure 17-a, a dissection of the slope into several levels of flat sections.

He then added a rectangle to his sketch, as in Figure 17-b, as a "local experiment" of putting a class-room building on the second level. As a result, his sketch now shows that the roof of the building rises considerably from

[5]Although Schön's reports on this event is fairly detailed, he does not go so far as to report the exact order in which the parts of Quist's sketches were drawn. Thus, the following should be viewed as our best reconstruction of the event from what Schön wrote about it.

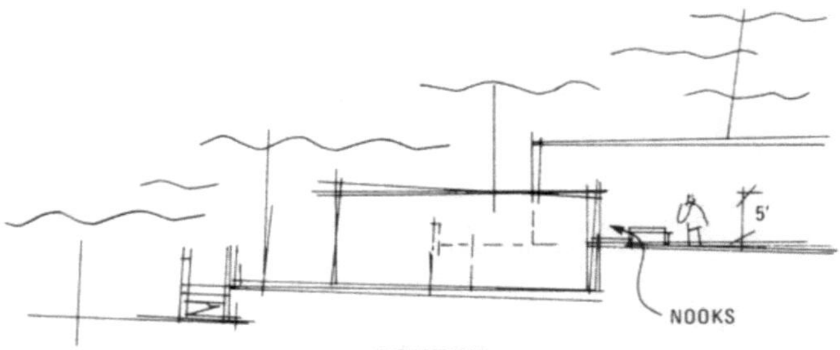

FIGURE 18 Actual sketch of Quist's, also reflecting the additional ideas added after the free ride. Reprinted with permission of Perseus Books Group, from Schön (1983); permission conveyed through Copyright Clearance Center, Inc.

the ground of the third level. This information came as an unexpected consequence, something automatically expressed as the result of expressing certain experimental ideas in sketch.

This idea then inspired the idea of creating a nook inside that second-level building, with the floor of the nook continuous to the ground on the third level. Because the rise of the roof is large enough, most children can comfortably stand inside the nook. Figure 18 shows the actual sketch of Quist's, also reflecting the additional information added after the free ride.

Lawson (1997) points to the free ride potentials of architectural sketches in the following way:

> Thus the drawing represents a sort of hypothesis or "what if" tool. With a plan, for example, the architect can say, what if the kitchen were here, the dining-room next to it and the living-room there? How could I then organize the entrance and the stairs? (Lawson 1997: p. 242.)

According to Goldschmidt (1994), "One reads off the sketch more information than was invested in its making" (p. 164), and such a reading-off is an essential step of what she describes as "interactive imagery" in design. Suwa et al. (2000) emphasizes that such a reading-off is frequently an "unexpected discovery" and that it can lead to the *invention* of new design requirements on the designer's part as opposed to his or her client's part. Thus, the line of research cited here abundantly demonstrates the importance of the free ride potentials of sketch drawings in architectural design.

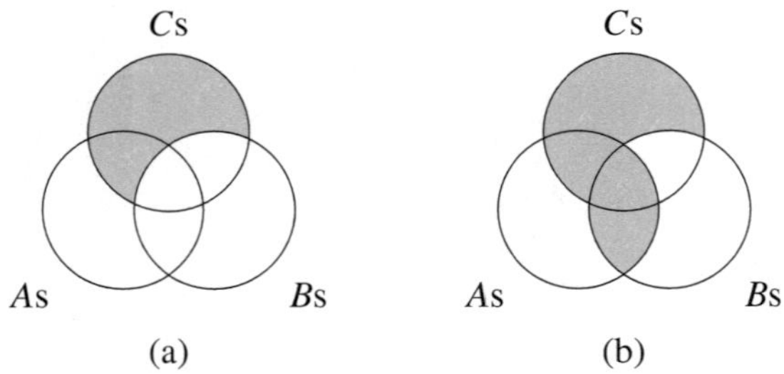

FIGURE 19 Venn diagrams supporting free rides: (a) before information θ_{15} is expressed, and (b) after it is expressed.

2.1.6 Further Examples: Venn Diagrams

There are also much simpler cases in which the phenomenon of free ride manifests itself. Consider the case where we use Venn diagrams to check the validity of the following syllogism:

(θ_{14}) All Cs are Bs.

(θ_{15}) No Bs are As.

(θ_{16}) (Therefore) no Cs are As.

We start with drawing three circles, labeled "As," "Bs," and "Cs," respectively. On the basis of the premises θ_{14} and θ_{15} of the syllogism, we shade the complement of the "Bs" circle with respect to the "Cs" circle (Figure 19-a) and then shade the intersection of the "Bs" circle and the "As" circle (Figure 19-b). Observing that the intersection of the "Cs" circle and the "As" circle is shaded as a result, we read off the conclusion θ_{16} of the syllogism, and decide that the syllogism is valid. This is clearly a case of free ride.[6]

2.1.7 Further Examples: Euler Diagrams

We use Euler circles to see if θ_{19} follows from θ_{17} and θ_{18}:

(θ_{17}) $C \subset B$

(θ_{18}) $B \cap A = \emptyset$

(θ_{19}) (Therefore) $C \cap A = \emptyset$

On the basis of the premises θ_{17} and θ_{18}, we draw a circle labeled "C" inside a circle labeled "B" (Figure 20-a), and then draw a circle labeled "A"

[6]For those who are not familiar with the semantics of Venn diagrams, shading generally means the emptiness of the set represented by the shaded area. So, for example, the shading in the complement of the "Bs" circle with respect to the "Cs" circle means that there are no Cs that are not Bs, i.e., that all Cs are Bs. Also, the shading in the intersection of the "Bs" circle and the "As" circle means that there are no such things that are both Bs and As, i.e., that no Bs are As (or equivalently, that no As are Bs).

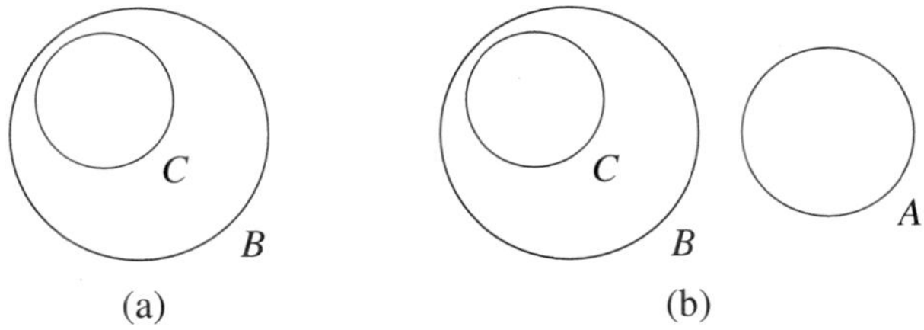

FIGURE 20 Euler diagrams supporting free rides: (a) before information θ_{18} is expressed, and (b) after it is expressed.

completely outside the "*B*" circle (Figure 20-b).[7] We observe that the "*C*" circle and the "*A*" circle do not overlap, and read off the conclusion, θ_{19}. Here again we see a case of free ride: expressing θ_{17} and θ_{18} in the diagram results in the expression of the other, consequential information, θ_{19}.

2.2 General Framework for Analysis

We thus have visited several places where the phenomenon of a free ride is discussed or suggested. Before embarking on the analysis of the phenomenon, we clarify a few key assumptions that underly all the analyses presented in this book. They are mainly concerned with the way a representation expresses information about a particular object or situation, and have their origin in situation semantics and its derivatives (Barwise and Perry 1983/1999; Barwise and Etchemendy 1987; Barwise and Seligman 1997). We, however, will make our exposition simple without going deeply into the details of the original model.

Figure 21 lays out our general picture on the way a representation expresses information about a particular object or situation. It shows four components: representation s, source type σ, represented object t, and target type θ. It also shows the relations, such as $\Rightarrow$ and $\rightsquigarrow$, holding between pairs of these components. We explain these components and relations one by one.

2.2.1 Representations and Source Types

Representations are such things as an individual diagram drawn on a particular sheet of paper, a map printed on a particular page of a brochure, and a sentence displayed on a particular computer monitor. Typically, a particular arrangement of symbols and colors appear in a representation, but a representation is not to be identified with its appearance. For example, when one

[7]Throughout this book, we assume the system of Euler diagrams, formalized by Hammer (1995) , where the non-emptiness of a region indicates the non-emptiness of the corresponding set. It differs from semantically weaker system of Euler diagrams, formalized by Hammer and Shin (1998) and Mineshima et al. (2012), where only the emptiness of a region is interpreted (to indicate the emptiness of the corresponding set.)

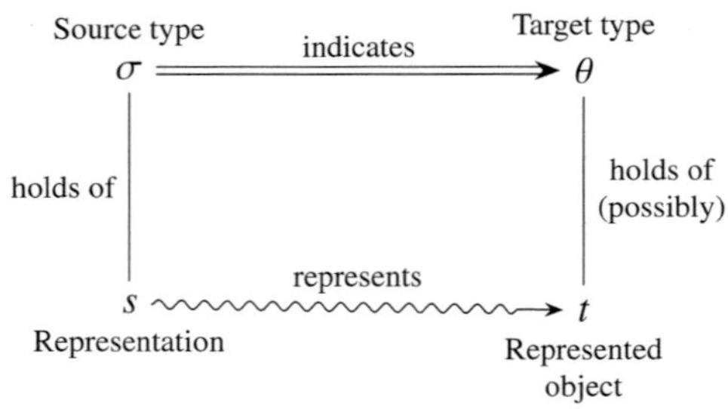

FIGURE 21 General framework for the analysis in this book.

draws exactly the same arrangement of symbols on two different occasions, the result is two different representations although they have exactly the same appearance. Similarly, two prints of *Downtown Chicago Map* published by the same map publisher are different representations even though they usually have exactly the same arrangement of symbols and colors.

In this sense, a representation is a *token*, while the arrangement of symbols or colors appearing on it is a *type*, or a property. We call a type or property that may hold of a representation a *source type*. Thus, given a particular print of *Downtown Chicago Map*, the following are source types that possibly hold of it:

(σ_1) There is a unique street line labeled "E Ontario."

(σ_2) A hotel symbol is at the upper-left corner of the intersection of the street lines labeled "N Rush" and "E Ontario."

(σ_3) A street line labeled "Chicago Av" is the fourth street line above a street line labeled "Lake" along a street line labeled "Oak Park Av."

These are what we sometimes call "state of affairs," and it is important to realize that they are types, rather than tokens, in the sense that they are properties that can hold of many different map tokens. In fact, they hold of all prints of the same edition of *Downtown Chicago Map*, with the exception of defective or damaged copies (changed colors, unprinted symbols, spilled coffee, etc.) These types even hold of revised maps of downtown Chicago, as long as they have a unique street line labeled "E Ontario", have a hotel symbol in an appropriate position, and have three street lines in an appropriate spatial relation. When a type holds of a token, we sometimes say that the token *supports* the type.

2.2.2 The Representation Relation

To explain the representing relation $\rightsquigarrow$ depicted in the bottom of Figure 21, we first need to introduce the concept of a representational act. A *representa-*

tional act is an act of a person, or a collaborative act of a group of people, that collects information about an object or situation and produces another object that expresses particular pieces of information about the first object. In the case of *Downtown Chicago Map*, some people collected information about a region in Chicago in a particular period of time, some designer for the map publisher used that information to design a map, and a technician printed the first print of the map. This sequence of acts makes the first representational act in this map-making practice. When the technician makes another print, this act makes a second representational act, and so on until the final printing act.

Note that each representational act in this sequence establishes a causal link from a region of Chicago in a particular period of time to a particular print of *Downtown Chicago Map*. Because of this causal link, that particular print is *about* that geographical region in the particular period, or it is a map *of* that particular region in that period. When a representation s is linked to a particular object t through a representational act, we say that *s represents t*, writing $s \rightsquigarrow t$.

Similarly, if the manager of a shop regularly hears from the part-time workers to decide on who will work on what day of the next week, and puts up a table of the next week's work schedule on the door of the office, a sequence of representational acts has arisen. Each act in this sequence establishes a causal link from a particular week to a particular work schedule table, making the latter a representation of the former. A particular work schedule table represents a particular week in this way.

2.2.3 Represented Objects and Target Types

Objects represented in this way are tokens rather than types, just as are representations. For example, a particular region of Chicago in a particular period of time is a token, of which various types can hold. We call types that may hold of represented objects *target types*. Thus, target types in the representational practice with *Downtown Chicago Maps* include:

(θ_{20}) There is a unique street named "E Ontario."

(θ_{21}) A hotel building is at the north-west corner of the crossing of streets named "North Rush" and "East Ontario."

(θ_{22}) A street named "Chicago Avenue" is the fourth street to the north of a street named "Lake" along a street named "Oak Park Avenue."

The particular region of Chicago in the particular period represented by a *Downtown Chicago Map* supports each of these target types. Note the difference from types σ_1–σ_3. Types θ_{20}–θ_{22} refer to possible arrangements of streets and buildings in a region in Chicago, rather than possible arrangements of lines and symbols on a sheet of paper.

2.2.4 The Indication Relation

The indication relation $\Rightarrow$ depicted in the upper part of Figure 21 is a relation from a source type to a target type. This relation is established by the semantic conventions for the relevant representational practice. A semantic convention is essentially arbitrary in its origin, but once people start conforming to it and believe that everybody conforms to it, there is a significant mutual benefit for them to keep conforming to it. It thus becomes a "self-perpetuating" constraint over the representational acts of a group of people.[8]

In the case of the representational practice with *Downtown Chicago Maps*, the designer of the map made an effort to draw a unique street line labeled "E Ontario" only when the information he or she had obtained ascertained that there was a unique street named "East Ontario" in the region of Chicago represented by the map. Partly because of this effort on the part of the designer, and partly because of the fidelity of the printing technician's work, the following regularity holds on the map-making practice with *Downtown Chicago Maps*:

(1) If there is a unique road line labeled "E Ontario" on the print, there is a unique road named "East Ontario" in the represented region.

The indication relation $\Rightarrow$ is intended to capture these kinds of constraints from source types to target types that arise from semantic conventions. Thus, we can rewrite (1) as $\sigma_1 \Rightarrow \theta_{20}$, referring back to the specifications of source types and target types in Subsections 2.2.1 and 2.2.3 above. In fact, $\sigma_2 \Rightarrow \theta_{21}$ and $\sigma_3 \Rightarrow \theta_{22}$ also hold, as the reader can easily verify.

To illustrate the relation $\Rightarrow$ with one more example, let us consider the case of work schedule tables. The manager draws a tick mark in the intersection of the "Ron"-row and the "Wed"-column only when he or she is sure that Ron will work on Wednesday in the week represented by the table. Conforming to this regularity serves for the benefit of both the manager and the workers, making it a self-perpetuating regularity over the representational practice with work schedule tables for this shop. In other words, the following is a semantic convention for this representational practice:

(2) If a tick mark appears in the intersection of a "Ron"-row and a "Wed"-column in the produced table, then Ron will work on Wednesday in the represented week.

So, we write $\sigma_4 \Rightarrow \theta_{23}$ where σ_4 and θ_{23} are defined in the following way:

[8] The interested reader might consult the general theory of conventions developed by the philosopher David Lewis Lewis (1975/1985), where he explains how an essentially arbitrary tendency can become a self-perpetuation regularity in human conduct. Particularly relevant to our discussion of semantic conventions here is his theory on how the tendency towards truthfulness and trustfulness develops into a regularity of conducts among a language's users.

(σ_4) A tick mark appears in the intersection of a "Ron"-row and a "Wed"-column.

(θ_{23}) Ron works on Wednesday.

In our analysis, we often deals with a case where a set Γ of source types and a set Δ of target types stand in a one-to-one correspondence under the indication relation $\Rightarrow$, i.e., the case where every source type in Γ indicates a unique target type in Δ and every target types in Δ is indicated by a unique source type in Γ. Under this condition, we say Γ *collectively indicates* Δ, and write $\Gamma \Rightarrow \Delta$. Here, the symbol $\Rightarrow$ is slightly abused to cover the relation between sets of types as well as the relation between individual types. Thus, the set $\{\sigma_1, \sigma_2, \sigma_3\}$ of source types in the system of *Downtown Chicago Maps* collectively indicates the set $\{\theta_{20}, \theta_{21}, \theta_{22}\}$ of target types, and we write $\{\sigma_1, \sigma_2, \sigma_3\} \Rightarrow \{\theta_{20}, \theta_{21}, \theta_{22}\}$. Intuitively, that Γ collectively indicates Δ by $\Rightarrow$ means that Γ is a minimal set of source types that is necessary to indicate all types in Δ.

2.2.5 Two-Tier Semantics

Figures 22-a and 22-b show the results of applying our general framework to some of the examples discussed above. The conditions shown in Figure 22-a let us say that particular print s of *Downtown Chicago Map* represents particular region t of Chicago in a particular period as being of type θ_{20}, i.e., as having a unique street named "East Ontario." Note that target type θ_{20} may or may not actually hold of region t. Although $\sigma_1 \Rightarrow \theta_{20}$ is a fairly stable regularity in the relevant map-making practice, it can fail in certain circumstances, due to errors of the map designer or other incidents. In such a case, θ_{20} does not hold of t even when s represents t as being of θ_{20}.

Whether θ_{20} may actually hold of t or not, we call θ_{20} a *semantic content of* particular map print s. Generally, a target type θ is a semantic content of a representation s if there is a source type σ such that σ holds of s and $\sigma \Rightarrow \theta$. We say s is an *accurate representation of t* if all the semantic contents of s actually hold of t.

Similarly, the conditions in Figure 22-b let us say that work schedule table s on a particular piece of paper represents particular week t as being of type θ_{23}, i.e., having Ron working on Wednesday. Target type θ_{23} is one of the semantic contents of table s, and s can be said to be an accurate representation of week t if every such semantic content of s holds of week t.

Generally, we characterize the semantic relation between a representation and a represented object in the following way:

Definition 1 (Representing As) Let θ be a target type. A representation s *represents* an object or situation t *as being of type* θ if (1) s represents t and (2) s supports a source type σ that indicates θ.

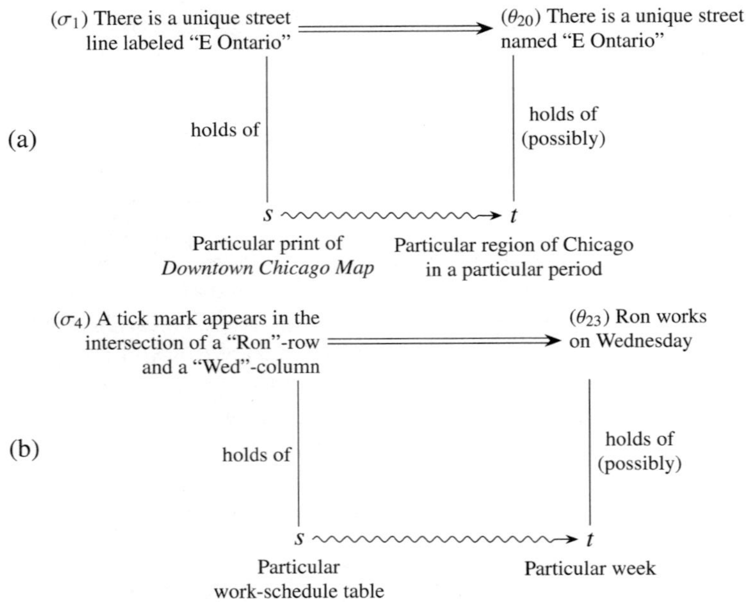

FIGURE 22 The general framework for analysis applied to the case of *Downtown Chicago Maps* and scheduling tables.

Our conception of the semantic relation is "two-tiered" in that it posits two relations, one being the representation relation $\rightsquigarrow$ from a representation to a represented object and the other being the indication relation $\Rightarrow$ from a source type to a target type. Two-tier semantics is implicit in Barwise and Perry's theory of linguistic meaning, also known as situation semantics (Barwise and Perry 1983/1999), and is more explicitly employed in Barwise and Etchemendy's analysis of the liar paradox (Barwise and Etchemendy 1987). Cummins' analysis of the contents of mental representations (Cummins 1996) also employs a version of two-tier semantics. Its most general formulation is found in Barwise and Seligman's model of representation systems based on their channel theory (Barwise and Seligman 1997; Shimojima and Barker-Plummer 2014) . Although this is not an appropriate place to explicate various arguments for this general view of semantic relation, we will pick out two of its advantages that are the most relevant to the analyses in this book.

First, because two-tier semantics clearly separates the type-level semantic relation $\Rightarrow$ from the token-level semantic relation $\rightsquigarrow$, we can focus our analysis on the former without being distracted by the issues related to the latter. For example, the concept of consequence tracking that we are about to introduce in this chapter is defined exclusively at the level of source types, target types, and the indication relation $\Rightarrow$ between them, and its ability to explain the free ride phenomenon comes from this fact. It would lose its conceptual coherence, and its explanatory force along with it, were it to be understood as being defined on representations, represented objects, and the semantic relation $\rightsquigarrow$ between them. To be sure, the overall account must come down to references to these token-level objects and relations in the end, but consequence tracking must be strictly understood to be a type-level relation. The same goes for all the other semantic properties introduced in this book. They are all defined at the type level, and to keep their status as type-level properties clear, we need a framework that does justice to both type-level and token-level objects and relations.

The other advantage of two-tier semantics relevant to our analysis is that it can naturally handle those representations that are not concerned with particular objects or situations in the world. According to the definition given above, a representation s can have a semantic content even if it does not represent any object or situation. Supporting a source type σ that indicates a target type θ is all that is necessary, and θ is counted as one of its semantic contents. Looking back at Figure 22-b, for example, target type θ_{23} is a semantic content of work schedule table s even if there is no particular week t that s represents. Maybe the manager has produced table s not to express information about any particular week, but only to consider one of various work schedules that the manager can possibly adopt in the future. Still, the arrangement of symbols on it is not meaningless. For one thing, a tick mark appearing in the intersection of the "Ron" row and the "Wed" column indicates that Ron works on Wednesday, and because of such a semantic content, table s serves for the intellectual activity of exploring possible work schedules.

Or take the example of design sketches discussed in Subsection 2.1.5. Since the architectural sketches were drawn for a design practice, they were not concerned with any particular land on the Earth, past or future. Nevertheless, they expressed certain pieces of information, and thus helped Quist and Petra maintain design ideas and discover their consequences. Or consider illustrations in a fictional story, or a diagram in a science magazine that depicts a general mechanism or phenomenon. All these are not concerned with any particular objects or situations in the world, but are still meaningful. Two-tier semantics does justice to these representations by separating the meaning relation $\Rightarrow$ at the type-level from the meaning relation $\rightsquigarrow$ at the token level. A semantic content of a representation is guaranteed by $\Rightarrow$ and survives even when $\rightsquigarrow$ fails.

$$\left\{\begin{array}{ll}\text{Ron works} & \text{Bob works}\\ \text{on Friday,} & \text{on Friday}\end{array}\right\} \vdash \left\{\begin{array}{ll}\text{Ken does} & \text{Jon does}\\ \text{not work on} & \text{not work on}\\ \text{Friday,} & \text{Friday}\end{array}\right\}$$

$$\left\{\begin{array}{ll}\text{Ron works} & \text{Ron does}\\ \text{at least} & \text{not work on}\\ \text{three days,} & \text{Friday}\end{array}\right\} \vdash \left\{\begin{array}{ll}\text{Ron works} & \text{Ron works}\\ \text{on Monday,} & \text{on Tuesday}\end{array}\right\}$$

FIGURE 23 Examples of constraints holding on the working weeks of a shop.

2.2.6 Constraints

Although it is not directly related to the semantic framework as it is described above, it is useful to explicitly introduce the notion of *constraint* here. A constraint is a regularity, of whatever origin or origins, that governs the distribution of types in a particular class of tokens. For example, it may mandate a token to support a particular type γ_4 if it supports a particular set of types, say $\{\gamma_1, \gamma_2, \gamma_3\}$. It may prohibit a token from supporting a particular set of types at once. It may mandate a token to always support a set of types.

Let Γ and Δ be sets of types. We say that there is a constraint from Γ to Δ, and write $\Gamma \vdash \Delta$, if *some* type in Δ must hold of a token if *all* types in Γ hold of that token. Thus, whenever we discuss a constraint $\Gamma \vdash \Delta$, set Γ is taken conjunctively and set Δ is taken disjunctively. Thus, for example, $\{\gamma_1, \gamma_2, \gamma_3\} \vdash \{\delta_1, \delta_2, \delta_3\}$ means that if γ_1 *and* γ_2 *and* γ_3 hold of a token, *either* δ_1 *or* δ_2 *or* δ_3 must hold of that token. Set Γ in the constraint $\Gamma \vdash \Delta$ is called the *antecedent* of the constraint, while set Δ is called its *consequent*.[9]

We apply this idea of constraint to both the sets of source types and the sets of target types in a representation system. The system of schedule tables mentioned in Subsection 2.2.3 gives a good example. Suppose the shop has four workers, Ron, Ken, Bob, and Jon, and exactly three workers work each day Monday through Friday. This regulation, along with the shop's limited personnel, make the constraints shown in Figure 23 hold on any working week of the shop. Remember that the antecedent of each constraint is taken conjunctively while the consequent is taken disjunctively.

---

[9]Formally, we require the set of constraints holding on a given set of tokens to be closed under the following rules:

Identity : $\{\alpha\} \vdash \{\alpha\}$ for every $\alpha \in \mathrm{typ}(\boldsymbol{A})$,

Weakening : If $\Gamma \vdash \Delta$, then $\Gamma \cup \Sigma_1 \vdash_{\mathcal{L}} \Delta \cup \Sigma_2$ for any $\Sigma_1, \Sigma_2 \subseteq \mathrm{typ}(\boldsymbol{A})$,

Global Cut : If there is a set $\Sigma \subseteq \mathrm{typ}(\boldsymbol{A})$ such that $\Sigma_1 \cup \Gamma \vdash \Delta \cup \Sigma_2$ for each partition $\langle \Sigma_1, \Sigma_2 \rangle$ of Σ, then $\Gamma \vdash_{\mathcal{L}} \Delta$.

Here, $\langle \Sigma_1, \Sigma_2 \rangle$ is a partition of Σ iff $\Sigma_1 \cup \Sigma_2 = \Sigma$ and $\Sigma_1 \cap \Sigma_2 = \emptyset$. This implies that the holding of constraints on a set of tokens follows the structural rules of classical logic. The mathematically inclined reader may want to consult Barwise and Seligman (1997), especially Chapters 9 and 12, for a formal development of the idea of constraint.

$$\left\{\begin{array}{ll}\text{Ron works} & \text{Jon works}\\ \text{at least} & \text{on four}\\ \text{three days,} & \text{days}\end{array}\right\} \vdash \begin{array}{l}\text{Ron and Jon work}\\ \text{together on at least two}\\ \text{days}\end{array}$$

FIGURE 24 Example of a consequential constraint holding on the working weeks of a shop.

$$\left\{\begin{array}{ll}\text{The ``Ron'' row} & \text{A cross mark}\\ \text{has at least} & \text{appears in the}\\ \text{three tick} & \text{intersection of}\\ \text{marks,} & \text{the ``Ron'' row}\\ & \text{and the ``Fri''}\\ & \text{column}\end{array}\right\} \vdash \left\{\begin{array}{ll}\text{A tick mark} & \text{A tick mark}\\ \text{appears in the} & \text{appears in the}\\ \text{intersection of} & \text{intersection of}\\ \text{the ``Ron'' row} & \text{the ``Ron'' row}\\ \text{and the ``Mon''} & \text{and the ``Tue''}\\ \text{column,} & \text{column}\end{array}\right\}$$

FIGURE 25 Examples of constraints holding on the work schedule tables.

What if the consequent Δ of a constraint $\Gamma \vdash \Delta$ is a unit set, say, $\{\delta\}$? Then $\Gamma \vdash \Delta$ means that if all members of Γ hold of a token, some member of Δ must hold of that token, i.e., type δ must hold of that token. Such a constraint is special in that it mandates the holding of a *definite* type δ, not *some* of several types, under the condition that all members of Γ hold. Thus, we give a special name to a constraint of the form $\Gamma \vdash \{\delta\}$, calling it a *consequential constraint*. For readability, we omit the curly brackets in the consequent of a consequential constraint, writing $\Gamma \vdash \delta$ instead of $\Gamma \vdash \{\delta\}$. Figure 24 gives an example of a consequential constraint holding on the working weeks of the shop that we have been discussing.

Our analyses requires the introduction of other kinds of constraints, such as *disjunctive constraints* or *inconsistencies*, and they will be introduced and explained as our analyses proceed.

Although the examples of constraints given here all involve target types, constraints involving source types of a representation system will be often referred to in our analyses. Assume that each table in the system of scheduling tables has 4×5 cells, where the four rows are labeled with the four workers, “Ron,” “Ken,” “Bob,” and “Jon,” and the five columns are labeled with the names of the days of the week, “Mon,” “Tue,” “Wed,” “Thu,” and “Fri.” Then the constraint shown in Figure 25 would hold on any table of this system.

The reader may have noticed the correspondence of this constraint to one of the constraints shown in Figure 23. It is in fact this kind of correspondence between constraints involving source types and those involving target types that account for many characteristic cognitive potentials of diagrammatic representation systems.

2.3 Analysis

2.3.1 Consequence Tracking

We are now ready to clarify the semantic property underlying a free ride potential of a representation system. Simply stated, a representation system has a free ride potential when a constraint governing representations *tracks* a consequential constraint governing represented objects. We will shortly explain what exactly this means, but it should be already clear that this idea has been anticipated in the passage of Barwise and Etchemendy we cited in Section 2.1.2. They point out (for the case of diagrams) that the generation of free information in a representation is due to the fact that we "choose a representational scheme appropriately, so that the *constraints on the diagrams* have a good match with the constraints on the described situation" (p. 22, italics added). Also, Lindsay (1988/1995) refers to "constraint propagation" or "constraint satisfaction" that "forces the display of conclusions" as the underlying mechanism of what he calls "non-deductive inference procedures" (p. 126). The following analysis is a clarification of these somewhat sketchy account of the process of a free ride.

The Case of Venn Diagrams

In our example involving Venn diagrams, we express the following premises θ_{14} and θ_{15} of the syllogism with a Venn diagram.

(θ_{14}) All *C*s are *B*s.

(θ_{15}) No *B*s are *A*s.

Let us look closely at what exactly occurs *on the part of the diagrams* during this operation. In order for a Venn diagram to express that information, it must have certain properties that correspond to that information. Namely, certain source types that indicate that information must hold of the Venn diagram. What are they? According to the semantic rules associated with Venn diagrams[10], a Venn diagram expresses information θ_{14} and θ_{15} if it supports the following source types:

(σ_5) The complement of a circle labeled "*B*s" with respect to a circle labeled "*C*s" is shaded.

(σ_6) The intersection of the circle labeled "*B*s" and a circle labeled "*A*s" is shaded.

In fact, the operations that we take in the example are ones that realize

[10]Shin (1991) has developed a logical system, with a model-theoretic semantics, that formalizes one of the standard ways in which we use Venn diagrams. Our example is based on the semantic rules for Venn diagrams formalized in her paper. Shin's book (1994) contains a more thorough exposition of her formalization, including a formal examination of the degree of expressiveness of a system of Venn diagrams.

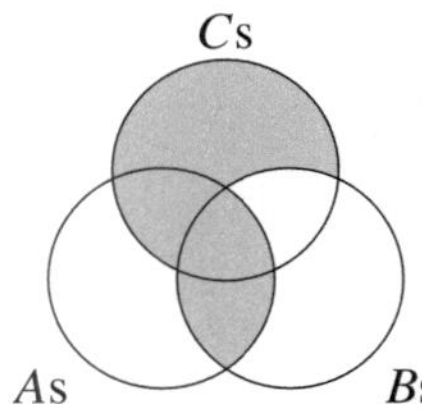

FIGURE 26 Venn diagram supporting a free ride from the premise $\{\theta_{14}, \theta_{15}\}$ to conclusion θ_{16}. Shown again.

these source types in a Venn diagram. Figure 26 shows, again, the Venn diagram that results from these operations.

Now let us ask ourselves. Is there anything that follows, *on the part of the diagram*, from these two facts? Well, as a matter of geometrical constraint, whenever source types σ_5 and σ_6 hold of a Venn diagram, the following source type *must* hold of the Venn diagram:

(σ_7) The intersection of a circle labeled "*C*s" and a circle labeled "*A*s" is shaded.

Note that this fact is a "side effect" of the operation that we have taken: the operation was taken in order to realize source types σ_5 and σ_6, not σ_7. Nevertheless, due to the semantic convention associated with Venn diagrams, this fact has an independent semantic value, namely, the information θ_{16}. We get information θ_{16} "for free," and decide that the syllogism is valid.

Figure 27 presents the mechanism of this "free ride" schematically. Since σ_5 and σ_6 indicate information θ_{14} and θ_{15}, respectively, due to our semantic conventions, one way of expressing premises θ_{14} and θ_{15} is to realize source types σ_5 and σ_6 in our Venn diagram. Now, as a matter of the structural constraints on Venn diagrams, the set of source types $\{\sigma_5, \sigma_6\}$ entails source type σ_7, which in turn indicates information θ_{16}. Thus, to check the validity of the syllogism, we do not have to infer conclusion θ_{16} from assumptions $\{\theta_{14}, \theta_{15}\}$ on our part. The constraint governing the Venn diagram takes over the work of making the necessary inference.

Note that this "inference" is *valid* indeed. Conclusion, θ_{16}, which we have obtained for free, is in fact a consequence of the premises $\{\theta_{14}, \theta_{15}\}$ that we started with. In other words, there is a constraint governing the relations between the sets of *A*s, *B*s, and *C*s, that makes θ_{16} a consequence of $\{\theta_{14}, \theta_{15}\}$. Thus, it is the match between this constraint on the target of the representation (the upper $\vdash$ in Figure 27) and the constraint on the representation (the lower $\vdash$) that guarantees our free ride's being valid. If it were not for this match, we simply could not count on the system of Venn diagrams for checking the validity of this syllogism.

{ (θ_{14}) All *C*s are *B*s,	(θ_{15}) No *B*s are *A*s }	$\vdash$	(θ_{16}) No *C*s are *A*s
$\Uparrow$	$\Uparrow$		$\Uparrow$
{ (σ_5) The complement of a circle labeled "*B*s" with respect to a circle labeled "*C*s" is shaded,	(σ_6) The intersection of the circle labeled "*B*s" and a circle labeled "*A*s" is shaded }	$\vdash$	(σ_7) The intersection of a circle labeled "*C*s" and a circle labeled "*A*s" is shaded

FIGURE 27 Semantic mechanism underlying a free-ride potential of Venn diagrams.

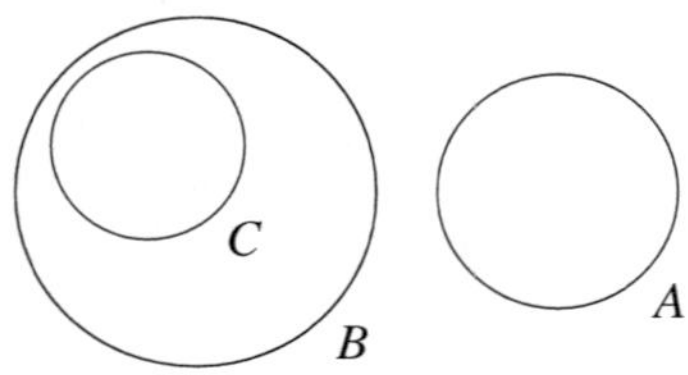

FIGURE 28 Euler diagram supporting a free ride from the premise $\{\theta_{17}, \theta_{18}\}$ to conclusion θ_{19}. Shown again.

The Case of Euler Diagrams

The case of Euler diagrams admits a similar analysis. We express the following premises in an Euler diagram:

(θ_{17}) $C \subset B$

(θ_{18}) $B \cap A = \emptyset$

We do this by realizing the following source types in an Euler diagram:

(σ_8) A circle labeled "*C*s" is inside a circle labeled "*B*s."

(σ_9) A circle labeled "*A*s" is outside a circle labeled "*B*s."

Figure 28 shows, again, the Euler diagram that results from this operation.

Note that these source types indicate the desired information, θ_{17} and θ_{17}, according to the semantic conventions we adopt for Euler diagrams. Now, as a matter of geometrical constraint governing Euler circles, these source types force the following source types to hold of the same diagram:

(σ_{10}) A circle labeled "*C*s" appears outside a circle labeled "*A*s."

Although the fact σ_{10} is a side effect of the operation we take, it has an independent semantic value, namely, the conclusion that $C \cap A = \emptyset$. We can read it off from the same Euler diagram, and again, we get a piece of information for free, without inferring it. Figure 29 presents the mechanism of this free ride.

$$\left\{ \begin{array}{cc} (\theta_{17}) & (\theta_{18}) \\ C \subset B, & B \cap A = \emptyset \end{array} \right\} \vdash \begin{array}{c} (\theta_{19}) \\ C \cap A = \emptyset \end{array}$$

$$\Uparrow \qquad\qquad \Uparrow \qquad\qquad \Uparrow$$

(σ_8) A circle labeled "*C*s" is inside a circle labeled "*B*s,"	(σ_9) A circle labeled "*A*s" is outside a circle labeled "*B*s"	$\vdash$	(σ_{10}) A circle labeled "*C*s" is outside a circle labeled "*A*s"

FIGURE 29 Semantic mechanism underlying the free-ride potential of Euler diagrams.

In both examples, we get a free ride from two premises to a single conclusion. The constraints responsible for the free rides are different in the two examples—the relevant constraint in 2.1 is one that governs the shadings of different areas of overlapping circles (the constraint from $\{\sigma_5, \sigma_6\}$ to σ_7), while the relevant constraint in 2.2 is one that governs the inclusion–exclusion relations between circles of different sizes (the constraint from $\{\sigma_8, \sigma_9\}$ to σ_{10}).

The structural constraints responsible for free rides are typically originated not only in these natural spatial constraints (geometrical and topological constraints) on the arrangements of symbols and colors in diagrams, but also in the syntactic stipulations adopted for the relevant representational practice with diagrams. For example, the constraint $\{\sigma_8, \sigma_9\} \vdash \sigma_{10}$ on Euler diagrams crucially depends on the syntactic stipulation that restricts the attachment of the same label ("*A*," "*B*," or "*C*") to only one circle. The constraint $\{\sigma_5, \sigma_6\} \vdash \sigma_7$ on Venn diagrams depends on a similar syntactic stipulation, as well as the so-called "partial-overlapping rule" (Shin 1994), which, roughly, mandates every circle appearing in a Venn diagram to overlap a part of every minimal region made up by the other circles. We use the neutral term "structural" to indicate this mixed nature of constraints governing the arrangements of symbols and colors in diagrams.

The General Case

At the core of our analysis is the observation that the phenomenon of free ride is attributable to a particular match between a constraint on representations and a constraint on their targets. We now describe, in general terms, this match of constraints responsible for a free-ride potential.

Definition 2 (Consequence Tracking) Let Δ be a set of target types and δ_i be a target type of a representation system $\mathcal{R}$. We say $\mathcal{R}$ *tracks a consequential constraint from* Δ *to* δ_i if there are a set Γ of source types and a source type γ_i such that:

1. $\Delta \vdash \delta_i$,

$$\begin{array}{ccc} \Delta & \vdash & \delta_i \\ \Uparrow & & \Uparrow \\ \Gamma & \vdash & \gamma_i \end{array}$$

FIGURE 30 Consequence tracking in a representation system.

2. $\Gamma \Rightarrow \Delta$,
3. $\gamma_i \Rightarrow \delta_i$,
4. $\Gamma \vdash \gamma_i$.

Figure 30 shows this general condition schematically. Recall that $\Gamma \Rightarrow \Delta$ is defined as the condition where Γ and Δ are in a one-to-one correspondence under $\Rightarrow$, so that realizing all members of Γ in a representation amounts to expressing all members of Δ.

Under this condition, system $\mathcal{R}$ has the potential for supplying a free ride in inference. By clause 2, one way of expressing Δ in system $\mathcal{R}$ is to realize Γ in a representation d of $\mathcal{R}$. By clause 4, this results in the realization of γ_i in the same representation, d. By clause 3, this means that representation d also expresses δ_i. And by clause 1, δ_i is a consequence of the set of information Δ that we started with. Thus, as stated in our characterization of a free ride potential, expressing a set of information Δ in a representation results in the expression of other, consequential information δ_i. This enables us to skip the mental deductive steps from Δ to δ_i, and to substitute them with the task of reading off δ_i from the representation.[11]

On this analysis, the utility of a free ride consists in the fact that it lets us express Δ in representation d in a way that saves us deductive steps from the assumptions Δ to the information δ_i (although the reasoner still has to read off δ_i from the representation). This process crucially depends on the existence of the structural constraint from Γ to γ_i holding on representation d and the constraint from Δ to δ_i holding on the represented object t. If it were not for the former, realizing the set of source types Γ in representation d would only result in Γ in d and nothing else that would indicate additional information. If it were not for the latter, the conclusion, δ_i, that we obtain for free in this process, might not hold of represented object t, even if the assumptions Δ did hold. We would get a ride, but it would be a ride to an invalid conclusion. We will come back to this point in Subsection 2.4.4.

[11]In a more formal development, we would require Γ in Definition 2 to be consistent ($\Gamma \nvdash \emptyset$) since an inconsistent Γ trivially entails an arbitrary γ_1 and the definition would comprise cases that support no actual free rides in the sense discussed in this chapter.

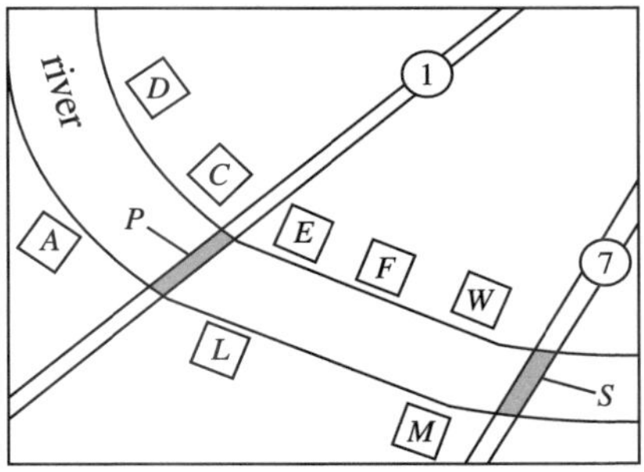

FIGURE 31 Harry's memory map, before information θ_1 was expressed. Shown again.

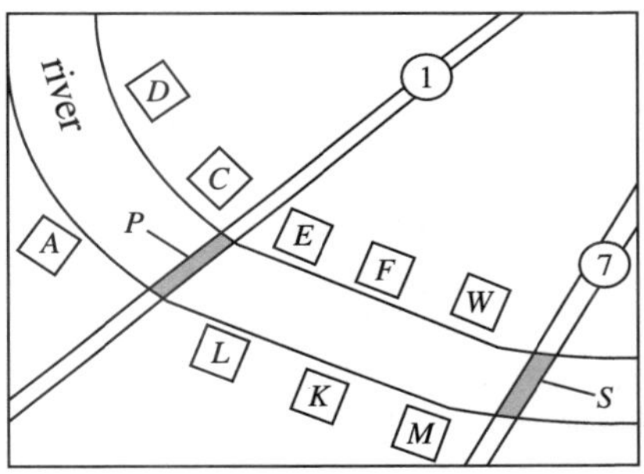

FIGURE 32 Harry's memory map, after information θ_1 was expressed. Shown again.

2.3.2 Application of the Notion to Specific Examples

Let us see how the notion of consequence tracking can be applied to analyze the specific cases of free ride discussed in the literature.

Memory Maps

We start with the example, given by Hohauser, where Harry updates a memory map (Figure 31) to another by adding information θ_1:

(θ_1) House *K* was halfway between houses *L* and *M* along the river.

The resulting map (Figure 32) "automatically" expresses various pieces of information, including θ_2, θ_3, θ_4, and θ_5 listed again below:

(θ_2) House *K* is across the river from house *F*.

(θ_3) House *K* is closer to road 1 than house *M* is.

(θ_4) House *M* is closer to bridge *S* than house *K* is.

(θ_5) House *K* and house *A* have road 1 in between.

Let us consider the expression of information θ_2. According to our general analysis, there must be some structural constraint on the map that is responsible for the automatic expression of this information. In fact, there is such a one. It is the constraint from:

(σ_{11}) Block F is halfway between blocks E and W along the band labeled "river."

(σ_{12}) The part of the band labeled "river" has straight parallel edges in the part between blocks E and W.

(σ_{13}) Block L is across the band labeled "river" from block E.

(σ_{14}) Block M is across the band labeled "river" from block W.

(σ_{15}) Block K is halfway between blocks L and M.

to:

(σ_{16}) Block K is across the band labeled "river" from block F.

The reader should convince themself that this is a reasonable constraint to assume to hold on the map. Note that source types $\sigma_{12}, \ldots, \sigma_{15}$ already hold of the initial map. Harry's operation preserves these source types while realizing σ_{15} in the resulting map. According to the semantic rules associated with this memory map, source types σ_{11}, σ_{12}, σ_{13}, σ_{14}, and σ_{15} indicate the following pieces of information respectively:

(θ_{33}) House F is halfway between houses E and W along the river.

(θ_{34}) The river has straight parallel edges in the part between houses E and W.

(θ_{35}) House L is across the river from house E.

(θ_{36}) House M is across the river from house W.

(θ_1) House K is halfway between houses L and M.

In our scenario, this much is a part of what Harry has already recollected about his home village. Now, due to the constraint specified above, the updated map also supports source types σ_{16}, which in turn indicates information θ_2. This information is not a part of what Harry has recollected, although it is a consequence of what he has. Thus, we obtain a free ride from the set of assumptions $\{\theta_{33}, \theta_{34}, \theta_{35}, \theta_{36}, \theta_1\}$ to information θ_2, exactly in the way that our general analysis of the phenomenon specifies. Here the structural constraint from $\{\sigma_{11}, \sigma_{13}, \sigma_{14}, \sigma_{12}, \sigma_{15}\}$ to σ_{16} on Harry's map matches with the constraint from $\{\theta_{33}, \theta_{35}, \theta_{36}, \theta_{34}, \theta_1\}$ to θ_2 via the indication relation. Figure 33 shows the consequence tracking property of Harry's map system responsible for this case of a free ride. Similar analyses apply to the free rides to the other pieces of information, including θ_3, θ_4, and θ_5.

$\left\{\begin{array}{l}(\theta_{33}) \text{ House } F \text{ is halfway between houses } E \text{ and } W \text{ along the river,} \\ (\theta_{34}) \text{ The river has straight parallel edges in the part between houses } E \text{ and } W, \\ (\theta_{35}) \text{ House } L \text{ is across the rvier from house } E, \\ (\theta_{36}) \text{ House } M \text{ is across the river from house } W, \\ (\theta_1) \text{ House } K \text{ is halfway between houses } L \text{ and } M\end{array}\right\} \vdash$ (θ_2) The house K was across the river from the house F

⇑ ⇑

$\left\{\begin{array}{l}(\sigma_{11}) \text{ Block } F \text{ is halfway between blocks } E \text{ and } W \text{ along the band labeled ``river,''} \\ (\sigma_{12}) \text{ The band labeled ``river'' has straight parallel edges in the part between blocks } E \text{ and } W, \\ (\sigma_{13}) \text{ Block } L \text{ is across the band labeled ``river'' from block } E, \\ (\sigma_{14}) \text{ Block } M \text{ is across the band labeled ``river'' from block } W, \\ (\sigma_{15}) \text{ Block } K \text{ is halfway between blocks } L \text{ and } M\end{array}\right\} \vdash$ (σ_{16}) Block K is across the band labeled "river" from block F

FIGURE 33 A consequence tracking property underlying the free ride potential of memory maps.

Thus, the general analysis shown in the last subsection is applicable to the process of a free ride based on Harry's maps. The analysis of this case highlights two interesting points about the phenomenon.

First, the above example shows that an operation that triggers a free ride can have a long sequence of operations before it, and its triggering a free ride may depend on the existence of this pre-sequence. Harry's placing block K between blocks L and M is just a sequel of the long sequence of operations that Harry takes to construct the map in Figure 31. Also, the operation triggers the free ride in question partly because it simply *preserves* some $(\sigma_{11}, \ldots, \sigma_{12})$ of the facts realized in the map by this pre-sequence. In terms of our general analysis (Figure 30), a triggering operation does not have to be an operation that creates all the pre-requisite facts Γ_1 for the free ride. Some or most of Γ_1 may have been originally created by operations before the triggering operation.

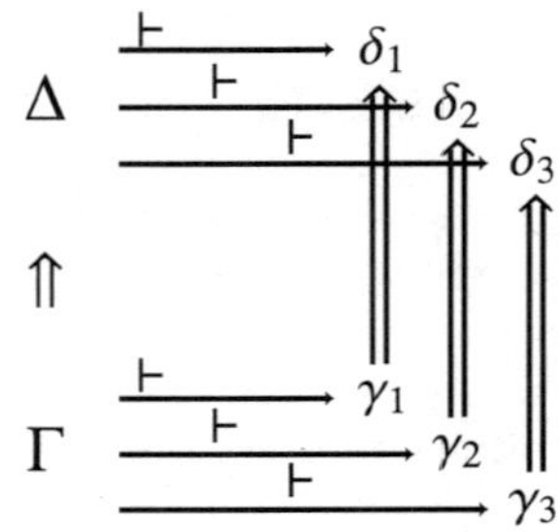

FIGURE 34 Case of multiple free rides.

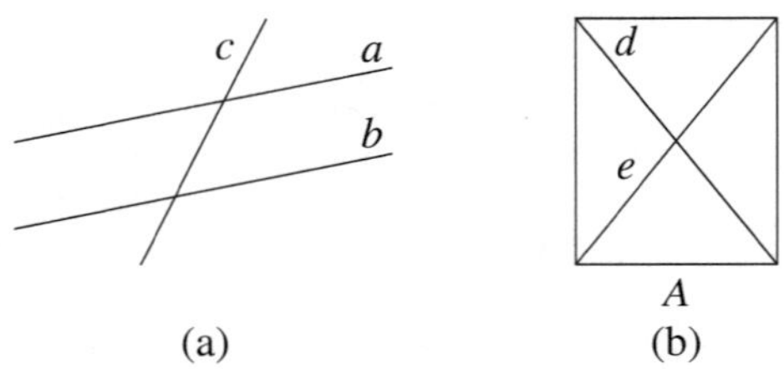

FIGURE 35 Geometry diagrams supporting free rides. Shown again.

Second, the example from Harry's memory maps shows that a single operation can trigger many free rides with different destinations. For example, Harry's placing block *K* between blocks *L* and *M* did not only trigger a free ride to θ_2, but also multiple free rides to θ_3, θ_4, θ_5, and other pieces of information. In terms of Figure 30, the symbol $\vdash$ that departs from Γ may have many siblings, as depicted in Figure 34. In fact, it is not hard to imagine that most of Harry's successive operations on memory maps trigger multiple free rides in similar ways. This seems to explain why the operations on memory maps have such an enormous computational advantage over pure thinking about the home village.

Geometry Diagrams

Larkin and Simon's examples are concerned with diagrams drawn in geometry proofs. We can think of such diagrams as depicting points, lines, and figures in the abstract space whose characteristics are defined by Hilbert's axiomatization of Euclidean geometry. For example, Figure 35-a expresses various pieces of information about three lines in Euclidean space. Among them are:

(θ_{38}) Lines *a* and *b* are parallel.

(θ_{39}) Line *c* cuts both *a* and *b*.

{ (θ_{38}) Lines *a* and *b* are parallel, (θ_{39}) Line *c* cuts both *a* and *b* }	$\vdash$	(θ_{40}) Lines *a*, *b*, and *c* make eight angles, four interior and four exterior
$\Uparrow$		$\Uparrow$
{ (σ_{17}) Line icons labeled "*a*" and "*b*" are approximately parallel, (σ_{18}) A line icon labeled "*c*" cuts both line icons labeled "*a*" and "*b*" }	$\vdash$	(σ_{19}) Line icons labeled "*a*," "*b*," and "*c*" make eight angles, four interior and four exterior

FIGURE 36 A consequence tracking property underlying the free ride potential of geometry diagrams.

What source types holding of the diagram indicate these pieces of information? They are:

(σ_{17}) Line icons labeled "*a*" and "*b*" are approximately parallel.

(σ_{18}) A line icon labeled "*c*" cuts both line icons labeled "*a*" and "*b*."

(By "line icons" we mean the black, thin, approximately straight marks on the paper that you are looking at. They must be distinguished from the lines in space that they stand for.)

Now, as a matter of structural constraint governing the diagram, source types σ_{17} and σ_{18} holding of the diagram yields the following state of affairs in the same diagram:

(σ_{19}) Line icons labeled "*a*," "*b*," and "*c*" make eight angles, four interior and four exterior.

This source type in turn indicates the following information.

(θ_{40}) Lines *a*, *b*, and *c* make eight angles, four interior and four exterior.

Thus, due to the structural constraint from $\{\sigma_{17}, \sigma_{18}\}$ to σ_{19}, we obtain a free ride from the assumptions $\{\theta_{38}, \theta_{39}\}$ to information θ_{40}. This free ride is again a correct one—there is a constraint on the Euclidean space that makes θ_{40} a consequence of $\{\theta_{38}, \theta_{39}\}$. Figure 36 summarizes our analysis.

A similar analysis applies to the other example provided by Larkin and Simon, where drawing a rectangle (labeled "*A*" in Figure 35-b) and its two diagonals (labeled "*d*" and "*e*") on paper generates the following information:

(θ_{41}) There is a point of intersection of segments *d* and *e*.

The structural constraint on diagrams responsible for this free ride is the one from source type:

(σ_{20}) Segment icons labeled "*d*" and "*e*" are the diagonals of a rectangular icon labeled "*A*."

to source type:

(σ_{21}) There is a point of intersection of segment icons labeled "*d*" and "*e*."

Note that σ_{21} is a source type holding of the diagram that we have drawn while θ_{41} is a piece of information about the Euclidean space that our diagram is about. We leave the reader to elaborate the analysis.

The reader may cast doubt on the utility of free rides in the case of geometry diagrams. Everybody who learns elementary geometry also learns to guard against so-called "accidental features": not everything that we can read off from a geometry diagram is a legitimate consequence of the assumptions that we have thrown in the diagram. So, even when a correct free ride occurs in drawing a geometry diagram, we may not be able to discern the free, legitimate information from other, possibly illegitimate pieces of information expressed in the diagram.

Nevertheless, it seems also true that those who study geometry at considerable length develop a certain tolerance to the accidental features, and can distinguish between the results of correct free rides and the accidental features. For example, many of us can "see" that information θ_{40} and θ_{41} expressed in Figures 35-a and 35-b are legitimate, while the following pieces of information expressed in the same diagrams are not:

(θ_{42}) Four of the eight angles defined by lines *a*, *b*, and *c* are acute.

(θ_{43}) None of the four angles defined by segments *d* and *e* are right angles.

Thus, as Larkin and Simon suggest, many of us do seem to be able to enjoy the utility of the free rides in spite of certain accidental features. In the next chapter, we will discuss how geometry diagrams and other diagrams come to exhibit accidental features in the first place. We will also suggest a strategy, called "semantic uplifting," that we seem to have developed to distinguish between the legitimate and the illegitimate from among the pieces of information presented in a single representation.

Other Systems

The analysis of free rides with Hyperproof diagrams (Subsubsection 2.1.2) is quite analogous to the one for the free rides with memory maps, and therefore is omitted. The analyses of the free rides with block drawings and design sketches (Subsubsections 2.1.3 and 2.1.5) are also similar, but the specification of the responsible consequence tracking properties is significantly more complex than other cases, so we record our analyses of these cases in Appendix A for the reader's scrutiny.

$$\begin{array}{ccc} \Delta & \vdash & \delta_i \\ \Uparrow & & \Uparrow \\ \Gamma & \vdash & \gamma_i \end{array}$$

FIGURE 37 A consequence tracking property of a representation system.

2.4 Questions and Responses

It is useful, at this point, to explicitly consider a couple of natural questions that our analysis might elicit. This will help clarify some of the most important points in the analysis, as well as its limitations.

2.4.1 Question: Who Does the Inference, Then?

Look back at your general analysis of free ride potentials, reproduced in Figure 37. You say that a free ride potential of the diagrams gives us an inference from a set Δ of premises to a conclusion δ_i *for free*. You want to say that it is a case of "inferential off-loading" (Scaife and Rogers 1996). But if somebody's load is taken off, somebody else needs to take it on. Miracles do not happen. Given Δ, what carries the inferential task of singling out the specific consequence δ_i from Δ?

Response

It is a structural constraint on the part of the diagrams. The free-ride potential gives us an inference from Δ to δ_i for free, and at the core of this potential is the existence of the structural constraint from Γ to γ_i. This constraint, having specific consequence γ_i, results from the combination of the spatial constraints on a plane and the syntactic stipulations adopted in the diagrammatic system. The major computation is done at this level, and the rest is the application of the semantic rule $\gamma_i \Rightarrow \delta_i$. So, if a miracle is there, it is the miracle of having such a set of syntactic and semantic conventions that exploit the spatial constraints on a plane to track the consequential constraint as shown in Figure 37.

Taking this point a little further, our analysis implies an interesting coupling of our cognitive functions with spatial constraint on real space. The physical operations of drawing a diagram let us project the premises of our inference onto an external diagram, exploit the spatial constraints holding there, and gain a "free ride" to a logical consequence. This is a paradigmatic case where cognitive processes are distributed over the external world (Hutchins 1995; Clark 1997; Clark and Chalmers 1998; Clark 2011).

There is an important point, however, in which the form of distributed cognition indicated by our analysis is different from what has been pointed out in the previous research on vision-based distributed cognition. According to O'Regan (1992), we do not have to construct an exact mental copy of our visual scene since the external world is generally quite stable and thus we can always return our attention to the relevant part of the visual scene and obtain the necessary information. Ballard et al. (1997) gave an empirical demonstration of this view, using the notion of deictic indexes to account for our frequent attention-returning behavior. In their second experiment, Spivey and Geng (2001) found that people could be so accustomed to the stability of an external display as to return attention to where the relevant visual information was, even when it was no longer available there. Thus, the previous findings on vision-based distributed cognition have been mainly concerned with how *memory* gains from interaction with the external environment. In contrast, our analysis is concerned with how *inference* gains from interaction with the external environment. In the case of a free ride, we take advantage of *spatial constraints* holding in real space to skip some inferential steps. It is in clear contrast to taking advantage of the *stability* of the external world to reduce the memory load.

2.4.2 Question: Is It Really Free?

Even if the burden of inference is carried by a structural constraint on the part of diagrams, there is still a cost for recognizing consequence γ_i that results from realizing Γ (see Figure 37). This cost can be significant, especially when the diagram is a complex one, such as the circuit diagram in Figure 38. You change the connection of a single pair of lines, and the diagram supports a whole lot of different source types concerning the connections and non-connections of the component icons. Some of these source types may be obvious, but others are significantly difficult to recognize, partly because of the difficulty of tracing the relevant lines in the clutter of other lines and symbols, and partly because of the difficulty of perceptually grouping the relevant symbols and lines into a meaningful macro structure.

Adding to this recognition cost is the cost of associating meaning δ_i to the recognized source type, γ_i. For this task, you need know the meaning relation $\gamma_i \Rightarrow \delta_i$ holds in the system, and be able to retrieve it upon the recognition of γ_i. For example, a pattern of connections in a group of part symbols may indicate a pattern of "line sync counter" (which I do not know exactly what it is), but remembering and applying this indication relation upon the recognition of the connection pattern in the diagram is no trivial task and may well take expertise. In such a case, the free ride is not at all free. It is a "cheap ride" at most (Gurr et al. 1998; Gurr 1999).

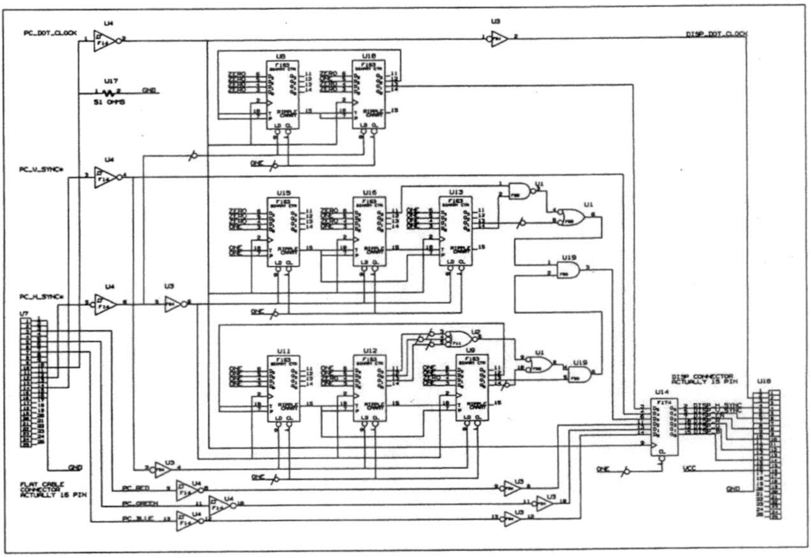

FIGURE 38 Electronic circuit diagram. Reprinted from Petre and Green (1992) with permission.

Reply

Granted, it is not totally free. All we claim is that the inferential part of the process is free. Because of the consequence tracking property in Figure 37, every diagram of the system expressing Δ by supporting Γ *expresses* δ_i automatically, but it does not mean that the user read off δ_i automatically. All the recognition costs and interpretation costs mentioned in the objection do occur. However, they are best considered separately from the cost of expressing a valid consequence in your diagram. In fact, the ability to recognize and interpret "global patterns" in a diagram has been another recurring theme in the literature, which we will discuss at length in Chapter 5.

2.4.3 Question: Is It Always Advantageous?

A free ride potential makes a diagram automatically express a consequence of a set of premises. The flip side of the coin is that it *forces* the diagram to express an additional piece of information, meaning that we have less discretion in what information to express in our diagram. So far as this additional piece is a consequence of the premises, there is nothing logically wrong about it, but for the purpose of controlling the information involved in our inferences, this feature of diagrams is disadvantageous. In fact, it is for the sake of this ideal of complete control that Gottlob Frege (1879/1967) invented the Begriffsschrift, the first formal language in history. He says:

> Its first purpose, therefore, is to provide us the most reliable test of the validity of a chain of inferences and to point out every presupposition that tries to sneak in unnoticed, so that its origin can be investigated (Frege 1879/1967: p. 6).

The free ride potential of diagrams raises the chance for certain pieces of information to "sneak in unnoticed," especially when the recognition-interpretation cost discussed in the previous objection is large. As we saw with the example of position diagrams in the introductory Chapter 1, linguistic representations, particularly those in a formal language, seem to be able to express only those pieces of information we consciously put into the individual sentences. Free rides are effectively prevented in such representation systems, and that can be an advantage.

Response

Granted, it is not advantageous for controlling the information involved in our inference. It is, however, advantageous for the purpose of making efficient inferences. Control and efficiency can be incompatible, and that explains why diagrams are often used with accompanying linguistic representations in the form of text or speech. The issue of control and efficiency, however, can be better discussed in the next chapter, where the discussion of the over-specificity potential of diagrams makes the problem of reduced control more evident for graphical representations.

2.4.4 Question: Possibility of Invalid Rides?

A free ride potential lets your diagram automatically express a piece of information as the result of your expressing a certain set of information. Is there any chance that the information so expressed is *not* a consequence of the set of information? As you explain it, a free ride is a fortunate case where the consequential constraint $\Gamma \vdash \gamma_i$ on representations matches with the consequential constraint $\Delta \vdash \delta_i$ on represented objects, as shown in Figure 39-a. This is a case of sound tracking of a consequential constraint. But what if the constraint $\Delta \vdash \delta_i$ fails, so that tracking of a consequential relation be invalid as shown in Figure 39-b? In such a diagrammatic system, whenever you express the premises Δ with the source types Γ, you end up expressing a non-consequential, and therefore misleading, information δ_i along with these premises. You would obtain a piece of information for free, but this ride would be invalid from the logical point of view. This could pose a serious problem to an inferential process.

Response

Yes, invalid tracking of a consequential constraint is a logical possibility. One can actually *imagine* a diagrammatic system with such a semantic property.

$$\begin{array}{ccc} \Delta & \vdash & \delta_i \\ \Uparrow & & \Uparrow \\ \Gamma & \vdash & \gamma_i \end{array} \qquad\qquad \begin{array}{ccc} \Delta & \nvdash & \delta_i \\ \Uparrow & & \Uparrow \\ \Gamma & \vdash & \gamma_i \end{array}$$

(a) (b)

FIGURE 39 Comparison of (a) valid and (b) invalid tracking of a consequential constraint.

If you use position diagrams cited in Chapter 1 to express non-transitive relations such as father-of, you will have an invalid free ride to the information that Mantle is the father of Moskowitz whenever you express that Mantle is the father of Mays and that Mays is the father of Moskowitz. Imagine that the left-of relation between names in Figure 40 indicates the father-of relation between the bearers of names. Such a system gives an invalid ride because it supports invalid tracking of a consequential constraint shown in Figure 41.

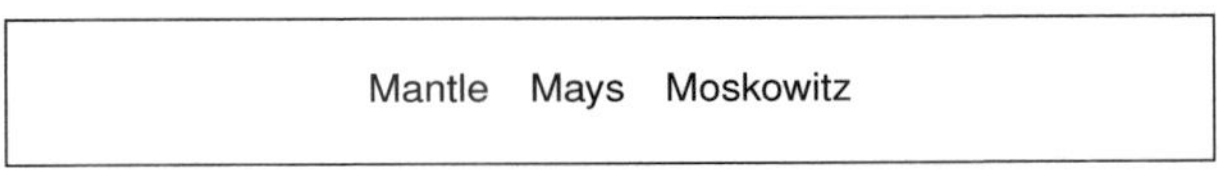

FIGURE 40 Position diagram where the left-of relation between names indicates the father-of relation between the bearers of the names.

However, diagrammatic systems with such a property seem to exist only in imagination. In real life, nobody wants to express the father-of relation with position diagrams. As far as I know, there is no real-life diagrammatic systems supporting this property and hence with a potential for invalid free ride. In fact, no diagrammatic systems cited in the rest of this book do unsound tracking of consequential constraints.

One possible explanation for this phenomenon would be that such a system is so blatantly misleading that none have survived the selection by human users. This may be broadly right, but it leaves most of important details unfilled just because we know next to nothing about how diagrammatic systems are invented and inherited with and without modifications. I conjecture that the phenomenon has to do with a more or less unified way in which many primitive diagrammatic systems are conceived. Elaboration of this account however is beyond the scope of this book.

2.5 Chapter Takeaways

- A free ride potential is the potential of a representation system in which

$$\begin{array}{ccccc}
\left\{ \begin{array}{c} \text{Mantle is the father} \\ \text{of Mays,} \end{array} \right. & \begin{array}{c} \text{Mays is the father} \\ \text{of Moskowitz} \end{array} & \left. \vphantom{\begin{array}{c} a \\ b \end{array}} \right\} & \nvdash & \begin{array}{c} \text{Mantle is the father} \\ \text{of Moskowitz} \end{array} \\
\Uparrow & \Uparrow & & & \Uparrow \\
\left\{ \begin{array}{c} \text{Mantle's name is to} \\ \text{the left of Mays'} \\ \text{name,} \end{array} \right. & \begin{array}{c} \text{Mays' name is to} \\ \text{the left of} \\ \text{Moskowitz's name} \end{array} & \left. \vphantom{\begin{array}{c} a \\ b \\ c \end{array}} \right\} & \vdash & \begin{array}{c} \text{Mantle's name is to} \\ \text{the left of} \\ \text{Moskowitz's name} \end{array}
\end{array}$$

FIGURE 41 Semantic mechanism underlying the invalid ride with imagined position diagrams.

expressing certain pieces of information in that system results in the expression of another piece of information that follows from them.

- Many diagrammatic representation systems have a free-ride potential, including systems of position diagrams, Venn diagrams, Euler diagrams, memory maps, block drawings, geometry diagrams, and design sketches.
- A free ride occurs when, due to constraints on spatial configurations of representing objects, a set of source types that collectively indicates a set of information entails a source type that indicates a valid consequence of the set of information.
- A free ride saves one the cost of a deductive inference to this valid consequence, but not the cost of recognizing and interpreting the source type that is automatically realized in one's diagram.

3

Potential of Over-Specificity

> Expressing a set of information in diagrams can mandate the selective expression of other, often non-consequential pieces of information. This forces us to either refrain from expressing that set of information in diagrams or tolerate the presence of the additional pieces of unwarranted information in our diagrams.

In this chapter, we will identify and analyze another central cognitive potential of diagrams that has been recurrently discussed in the literature. This is the potential of over-specificity briefly explained in Chapter 1—quoted above. While the free-ride potential discussed in the last chapter captures a largely desirable aspect of diagrams, this potential captures a largely undesirable aspect of them. The organization of this chapter is modeled on that of the last chapter. Section 3.1 explores various places in the literature where this potential is described or suggested, demonstrating that they point to a single common phenomenon. Section 3.2 illustrates and defines the semantic property that underlies the potentials of over-specificity in graphical systems, namely, *the projection of a disjunctive constraint*. We will then apply this concept to analyze selected examples cited in the previous studies. Section 3.3 is devoted to a discussion of some essential questions that need be addressed to clarify our analysis.

3.1 The Phenomenon

3.1.1 Hohauser (1982): Memory Maps

Let us look again at the case of Harry's memory maps, cited in Hohauser (1982). Recall that Harry constructed the memory map in Figure 42 on the basis of his somewhat fragmentary memory about his home village.

Harry now recalls the following piece of information about his village.

(θ_1) House B is somewhere between houses A and K along the river, without any overlap with roads and other houses.

Before forgetting it again, he wants to record this information in his map by

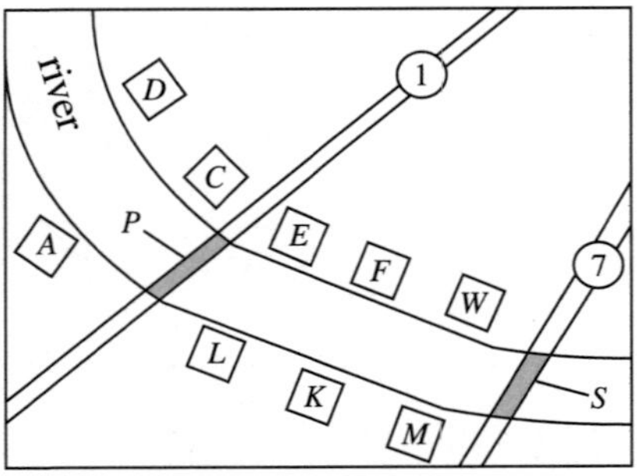

FIGURE 42 Harry's memory map that turns over-specific if information θ_1 is added.

placing the *B* block between the *A* and *K* blocks. However, there are more than one way of doing this. He could put the *B* block between the *A* block and road line 1, or between road line 1 and the *L* block, or between the *L* and *K* blocks (excluding the cases the *B* block comes on top of the other blocks and the road line). He cannot decide which alternative to take because he cannot remember which of the following was true:

(θ_2) House *B* is between house *A* and road 1.

(θ_3) House *B* is between road 1 and house *L*.

(θ_4) House *B* is between houses *L* and *K*.

Harry's problem occurs due to the over-specificity of his system of memory maps. Let Θ_m be the set of information already expressed in the map in Figure 42 (as we did previously, we assume that there were fixed rules of interpreting his memory maps). The problem is that Harry's memory maps cannot express the set of information $\Theta_m \cup \{\theta_1\}$ in isolation, without simultaneously expressing either θ_2, θ_3, or θ_4. In other words, any representation in the system of Harry's memory maps must be specific about the choice between θ_2, θ_3, and θ_4 if it is to express set of information $\Theta_m \cup \{\theta_1\}$. Stenning and Oberlander (1995) aptly characterized the "content specificity" of a representation system as "the demand by a system of representation that information in some class be specified in any interpretable representation" (p. 98). We borrow their idea, and say that the system of Harry's memory maps is *over-specific in expressing* $\Theta_m \cup \{\theta_1\}$.

Now, the efficacy of a representation system partly depends on its ability to express, in an isolating manner, an *arbitrary* chunk of information that has been obtained in the course of reasoning. In this respect, Harry's system of memory maps is ineffective due to over-specificity in expressing set of information $\Theta_m \cup \{\theta_1\}$. We can express $\Theta_m \cup \{\theta_1\}$ in a map, but cannot express it in a manner that abstracts away from alternatives θ_2, θ_3, and θ_4. In words of

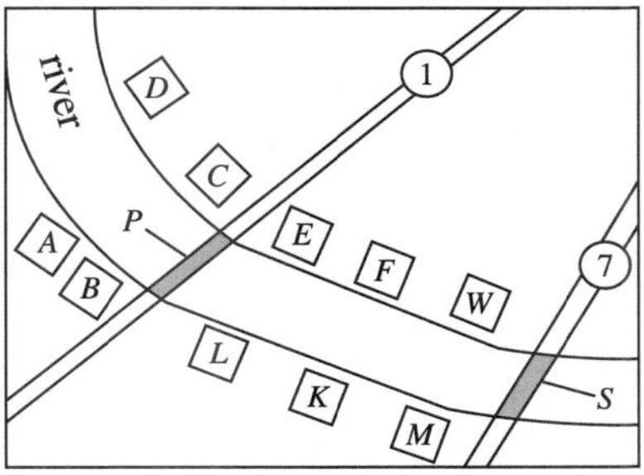

FIGURE 43 Harry's memory map, after information θ_1 is expressed with accidental feature σ_1.

Stenning, Oberlander , and other researchers, the system "limits abstraction" (Stenning and Oberlander 1995: p. 97), has a "particularity feature" (Kulpa 2003: p. 101), or allows "a smaller degree of discretion" about the selection of information to be expressed (Norman 2000: p. 110).

Note that this over-specificity of Harry's system can lead the representations to hold so-called "accidental features ." Although Harry actually knows better, let us suppose he went ahead and placed the *B* block between the *A* and *K* blocks, choosing a particular location, say, between the *A* block and road line 1. The map in Figure 43 would result. This map not only supports source types:

(σ_1) The *B* Block is between the *A* block and road line 1.

but also:

(σ_2) The *B* block faces the *C* block across the "river" band.

(σ_3) The *B* block is closer to bridge icon *P* than the *L* block is.

(σ_4) The *A* block is closer to the *B* block than the *L* block is.

Note that source types σ_2, σ_3, and σ_4 are all consequences of Harry's arbitrary choice of putting the *B* block between the *A* block and road line 1. Thus, σ_1, σ_2, σ_3, and σ_4 are *accidental features* of the map. Nevertheless, they respectively have the following independent semantic values:

(θ_2) House *B* is between house *A* and road 1.

(θ_6) House *B* is across the river from house *C*.

(θ_7) House *B* is closer to bridge *P* than house *L* is.

(θ_8) House *A* is closer to house *B* than house *L* is.

This is why the accidental features σ_1– σ_4 are potentially dangerous. In the course of his problem solving, Harry may forget that they are accidental features, and may base his future reasoning on the information indicated by

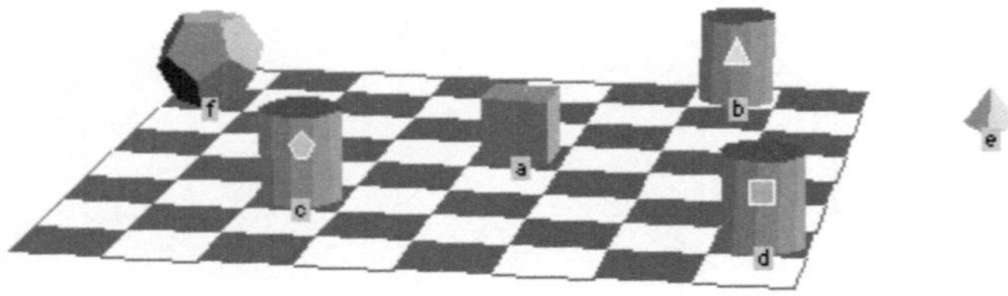

FIGURE 44 Hyperproof diagram that turns over-specific if the sentence `SameCol(e,a)` or `SameSize(b,f) ∨ SameSize(c,f)` is applied.

them. Harry's reasoning becomes invalid at this point, since none of θ_2, θ_6, θ_7, or θ_8 follow from the set of information $\Theta_m \cup \{\theta_1\}$ that Harry has accepted.[12]

3.1.2 Barwise & Etchemendy (1994): Hyperproof Diagrams

The system of Hyperproof diagrams in Barwise and Etchemendy (1994) exhibits certain over-specificity, as they themselves discussed (p. 22). We will use a slightly different example to illustrate the same point as theirs.

As we mentioned in the last chapter, the overall representation system that Hyperproof adopted was a heterogeneous system. It allowed interactions between diagrams and first-order sentences in a single proof. Suppose our premises are given both in the diagram in Figure 44 and in the following sentences:

- `Small(d)`
- `SameCol(e,a)`
- `SameSize(b,f) ∨ SameSize(c,f)`

Note that the icon labeled "*e*" is placed off the board icon in Figure 44. This is the Hyperproof's way of expressing that block *e* exists *somewhere* on the board, without specifying where exactly it is placed. Now, Hyperproof has a rule of inference, called "Apply," that allows us to incorporate the information content of a sentence into a diagram. How do we "Apply" the first sentence into the given diagram? Well, we can simply change the size of the cylinder icon labeled "*d*" to a small icon, which is easy and intuitive. What about the second sentence? We know that we should move the tetrahedron icon labeled "*e*" somewhere in the same column as the cube icon labeled "*a*." Unfortunately, we cannot decide on exactly where, and this is similarly so for the third sentence. We know that we should change either the cylinder icon

[12]This however does not mean that all the information that get expressed by placing the *B* block there is invalid. For example, the information that House *B* is on the opposite bank from house *C*, which gets expressed by this operation, is a consequence of $\Theta_m \cup \{\theta_1\}$. The same operation on a diagram can both trigger a free ride and causes accidental features. I thank Dave Barker-Plummer for pointing this out to me.

labeled "b" or the cylinder icon labeled "c" into the size of the dodecahedron icon labeled "f," but we do not know which icon to change.

This means that we cannot incorporate the information content of the second sentence or the third sentence into the given Hyperproof diagram, without making an arbitrary choice about the exact location of block e or the sizes of blocks b and c. To be more precise, let Θ_h be the information content of the given Hyperproof diagram, and θ_a, θ_b, and θ_c be the information content of the three sentences in that order. Then, our observation is that, while a Hyperproof diagram can express the information set $\Theta_h \cup \{\theta_a\}$ in isolation, no Hyperproof diagram can express $\Theta_h \cup \{\theta_b\}$ or $\Theta_h \cup \{\theta_c\}$ in isolation. The system of Hyperproof diagrams is over-specific in expressing these sets of information.

3.1.3 Barwise & Etchemendy (1991): Geometry Diagrams

The example of Harry's memory maps revealed that when a representation system exhibits over-specificity, there is a chance that representations in the system will support accidental features. Accidental features abound especially in geometry diagrams, and a student in a geometry class frequently appeals to an accidental feature of a diagram, and makes his or her proof invalid. Berkeley (1710) alludes to such a fallacy when he notes, "Having demonstrated that the three angles of an isosceles rectangular triangle are equal to two right ones, I cannot therefore conclude this affection agrees to all other triangles which have neither a right angle nor two equal sides" (Introduction, paragraph 16). Kulpa (2003) pointed out that Aristotle had already discussed accidental features of geometry diagrams:

> For there is in such activity an incidental affection identical with one also incidental in geometrical demonstrations. For in the latter case, though we do not for the purpose of the proof make any use of the fact that the quantity in the triangle [for example, which we have drawn] is determinate, we nevertheless draw it determinate in quantity. (Aristotle 350 BC/1941: p. 608, the note in parentheses by the translator, J. I. Beare.)

In the same vein, Barwise and Etchemendy (1991) discuss a geometry diagram (Figure 45) commonly used in a proof of the Pythagorean Theorem. Barwise and Etchemendy note, "if any piece of our reasoning had appealed to the observation that, in our diagram, a is greater than b, the proof would have been fallacious, or at any rate would not have been as general as the theorem demands"(p. 14). Similarly, Giaquinto (2007) summarizes the danger of geometry diagrams as consisting in the fact that "they tempt one to make unwarranted generalizations, as one's thinking may too easily depend in an unnoticed way on a feature represented in the diagram that is not common to all members of the class one is thinking about" (p. 77).

Now, the accidental features observed in these examples have their root in the over-specificity of the system of geometry diagrams. Berkeley's diagram

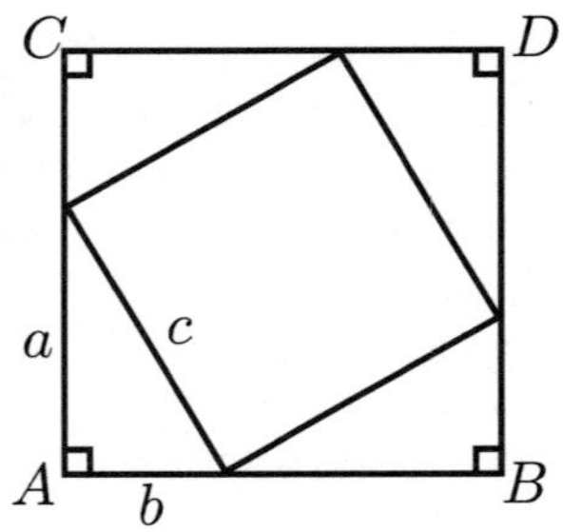

FIGURE 45 Geometry diagram used in a typical proof of the Pythagorean Theorem.

of a triangle has the features of being rectangular and equilateral *because* no geometry diagrams represent a figure as a triangle without representing it as having particular angles and particular sides. Of course, we do not have to represent it as a right isosceles triangle. However, if we are to draw a triangle at all, we have to make *some* choices about the sizes of its angles and sides. Likewise, the diagram in Barwise and Etchemendy's example did not have to represent the triangle as having side a longer than b. Yet, every geometry diagram has to make a choice anyway about the relative lengths of the triangle's three sides. The system of geometry diagrams is over-specific in representing a polygon as a right triangle.

Of course, these remarks are only true of the cases in which geometry diagrams are interpreted *on the semantic rules that a layman would adopt*. After several courses of high-school geometry, people seem to develop a sophisticated way of interpreting geometry diagrams, and learn to ignore the particular features that may mislead the proofs. Relative to such interpretation rules, the system of geometry diagrams is not over-specific as we described above. The topic is of great importance in revealing a way a "primitive" diagrammatic system can be modified to a more sophisticated system, and we will come back to it in Subsection 3.3.1. However, our purpose here is not to investigate such a form of system-change, but just to point out over-specificity based on unsophisticated, yet prevalent interpretations of geometry diagrams.

Aside: Philosophers on over-specificity

The topic of over-specificity has a long history in philosophy, although it is mainly discussed in connection to internal representations, rather than external representations.

We have mentioned that Aristotle had discussed the over-specificity of geometry diagrams. It was in the context of demonstrating that the presentation of a group of objects in our mind can have a determinate quantity ("incidental affection") when it is not essential to our thinking (Aristotle 350

BC/1941). In demonstrating why our mental representations cannot be "abstract" in Locke's sense, Berkeley (1710) claims that it is impossible to imagine a hand or an eye without imagining some particular shape or color of it; neither is it possible to frame the idea of a man without making him either black or white or tawny, straight or crooked, tall or low or middle-sized (paragraph 10). Likewise, we can never have the general idea of a triangle that is "neither oblique nor rectangle, equilateral nor scalenon, but all and none of these at once" (paragraph 13). Pursuing the same line of argument, Hume (1739) contends, "It is impossible to form an idea of an object that is possessed of quantity and quality, and yet is possessed of no precise degree of either" (Book 1, section VII). From these observations, Berkeley and Hume conclude that no mental representations are abstract.

Pylyshyn (1973) discusses the over-specificity of mental representations to reach an conclusion opposite to Berkeley's and Hume's. He thinks that there is an empirical evidence that mental representations are often *not* specific. According to him, "It would be quite permissible...to have a mental representation of two objects with a relationship between then such as 'beside,'" while "such a representation need not contain a more specific spatial-relation term such as 'to the left of' or 'to the right of'" (p. 11). On the basis of this observation, Pylyshyn argues that mental representations cannot be anything like images, because "it would seem to be an unreasonable use of the word 'image,' however, to speak of an image of two objects side by side without one of the relations between them being either 'to the left of' or 'to the right of.'" In the same spirit, Dennett (1969) argues:

> Consider the Tiger and his Stripes. I can dream, imagine or see a striped tiger, but must the tiger I experience have a particular number of stripes? If seeing or imagining is having a mental image, then the image of the tiger *must*—obeying the rules of images in general—reveal a definite number of stripes showing, and one should be able to pin this down with such questions as "more than ten?", "less then twenty?" (Dennett 1969: p. 136.)

Since Berkeley and Hume assume that all mental representations are image-like, they conclude that mental representations cannot be abstract (opposing to Locke). Since Dennett and Pylyshyn assume, as an empirical fact, that mental representations are often *un*specific in their contents, they conclude that they cannot be image-like. Although the occlusions are opposite, these philosophers share the common observation that some external representations, especially the ones called "pictures" or "images," are over-specific—they cannot present a certain combination of information in isolation, without thereby presenting additional pieces of information. If images and pictures did not have this property, then both arguments, either in the Berkeley-Hume line or in the Pylynshyn-Dennett line, would collapse.

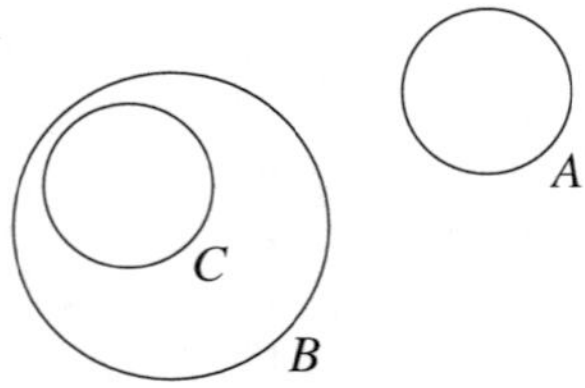

FIGURE 46 Euler diagram that expresses premises θ_9 and θ_{10}.

3.1.4 Further Examples: Euler Diagrams (Case 1)

Suppose we use Euler diagrams to see what conclusions can be deduced from the following premises:

(θ_9) $C \subset B$

(θ_{10}) $B \cap A = \emptyset$

(θ_{11}) $B \subset D$

We start by expressing the first two premises θ_9 and θ_{10} in an Euler diagram (Figure 46).

We now want to express premise θ_{11} in the diagram. According to a semantic convention we adopt for Euler diagrams, this requires us to enclose the circle labeled "*B*" in a circle labeled "*D*." However, such a circle must either have some overlap with the circle labeled "*A*" (Figure 47-a) or have no overlap with it (Figure 47-b).[13] Given our semantic convention, the two alternatives respectively indicate the following pieces of information:

(θ_{12}) $D \cap A \neq \emptyset$

(θ_{13}) $D \cap A = \emptyset$

Since neither follows from the given premises $\{\theta_9, \theta_{10}, \theta_{11}\}$, we cannot take the action required to express premise θ_{11}.

3.1.5 Further Example: Euler Diagrams (Case 2)

This time, we are given the following set of premises:

(θ_9) $C \subset B$

(θ_{10}) $B \cap A = \emptyset$

(θ_{16}) $A \subset D$

The first two premises are the same as those in the last example, so we start by drawing the diagram in Figure 46. We now want to express premise θ_{16}

[13] Actually, the first alternative has four sub-alternatives with different spatial relations between the "*D*" circle and the "*A*" circle. Figure 47-a is for the one of these sub-alternatives. However, our example would still make a case for over-specificity without listing these sub-alternatives.

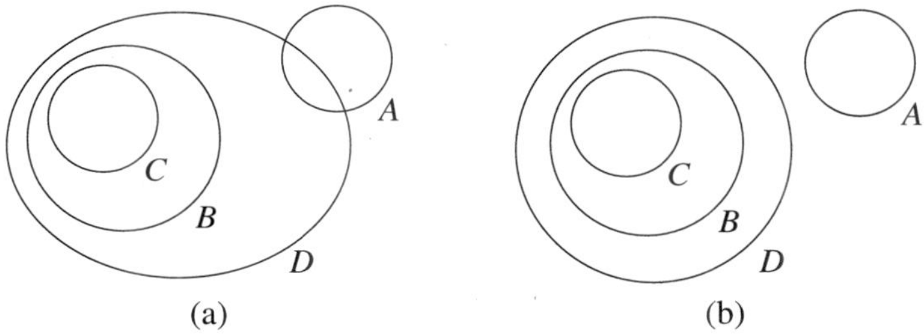

FIGURE 47 Alternative ways to add premise θ_{11} to Figure 46.

in the diagram. On our semantic convention, this requires us to enclose the "A" circle in a circle labeled "D." However, such a "D" circle must either have no overlap with the "B" circle or the "C" circle (Figure 48-a), or only overlap the "B" circle (Figure 48-b), or overlap both (Figure 48-c).[14] The three alternatives indicate the following pieces of information:

(θ_{17}) $D \cap B = \emptyset$ and $D \cap C = \emptyset$

(θ_{18}) $D \cap B \neq \emptyset$ and $D \cap C = \emptyset$

(θ_{19}) $D \cap B \neq \emptyset$ and $D \cap C \neq \emptyset$

Since none of these follow from the given premises $\{\theta_9, \theta_{10}, \theta_{16}\}$, we have to give up on adding premise θ_{16} to the diagram.

Both examples are instances of the phenomenon of over-specificity we are concerned with. The first example illustrates over-specificity of the system of Euler diagrams about alternatives θ_{12} and θ_{13} in expressing the set of information $\{\theta_9, \theta_{10}, \theta_{11}\}$. The second example illustrates over-specificity of the same system about alternatives θ_{17}, θ_{18}, and θ_{19} in expressing the set of information $\{\theta_9, \theta_{10}, \theta_{16}\}$. Our question is what imposes these particular kinds of over-specificity on the Euler diagrams, given the syntactic stipulations and semantic conventions we adopt for them.

3.2 Analysis

3.2.1 Projection of Disjunctive Constraints

Simply put, a representation system has an over-specificity potential when it projects what we will call "a disjunctive constraint" through the indication relation. We start by illustrating this account with the two examples of Euler diagrams just presented. Again, our strategy is to look closely at what happens to the diagrams during the operations in the examples.

[14] The second alternative has two sub-alternatives and the third alternative has six sub-alternatives. Again, it is not essential to list them separately for our purposes.

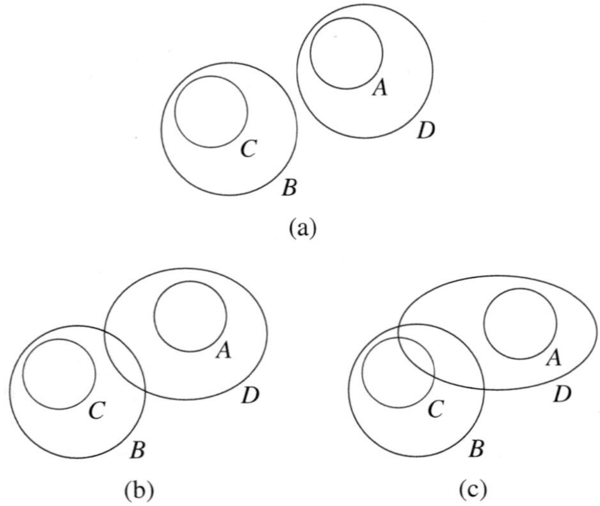

FIGURE 48 Alternative ways to add premise θ_{16} to Figure 46.

The Case of Euler Diagrams (Case 1)

In the first case of Euler diagrams, we first express premises θ_9 and θ_{10} in a diagram (Figure 46). Given our semantic convention for Euler diagrams, this means that we realize two source types in a diagram:

(σ_5) A circle labeled "C" is inside a circle labeled "B."

(σ_6) A circle labeled "A" is completely outside a circle labeled "B."

Now, to express assumption θ_{11}, we want to realize the following source type in the diagram (while preserving σ_5 and σ_6):

(σ_7) A circle labeled "D" encloses a circle labeled "B" in it.

The trouble is that doing so makes the diagram express either information θ_{12} or information θ_{13}. What does this mean in terms of the source types holding of the diagram? It means that whenever the diagram supports source type σ_7 while preserving σ_5 and σ_6, it *must* support either of the two source types:

(σ_8) A circle labeled "D" has no overlap with a circle labeled "A."

(σ_9) A circle labeled "D" overlaps a circle labeled "A."

In other words, there is a structural constraint governing the Euler diagram that makes the *disjunction* of σ_8 and σ_9 a consequence of $\{\sigma_5, \sigma_6, \sigma_7\}$. It is this structural constraint on Euler diagrams that gives rise to the particular over-specificity of the Euler system we have observed.

Figure 49 summarizes how this is so. Here, we realize source types $\{\sigma_5, \sigma_6,$

$$\left\{ \begin{array}{ll} (\theta_9) & C \subset B, \\ (\theta_{10}) & B \cap A = \emptyset, \\ (\theta_{11}) & B \subset D \end{array} \right\} \qquad \left\{ \begin{array}{ll} (\theta_{12}) & D \cap A \neq \emptyset, \\ (\theta_{13}) & D \cap A = \emptyset \end{array} \right\}$$

$$\Uparrow \qquad\qquad\qquad \Uparrow$$

$$\left\{ \begin{array}{ll} (\sigma_5) & \text{A circle labeled "}C\text{" is inside a circle labeled "}B\text{,"} \\ (\sigma_6) & \text{A circle labeled "}A\text{" is completely outside a circle labaled "}B\text{,"} \\ (\sigma_7) & \text{A circle labeled "}D\text{" encloses a circle labeled "}B\text{" in it} \end{array} \right\} \vdash \left\{ \begin{array}{ll} (\sigma_8) & \text{A circle labeled "}D\text{" has no overlap with a circle labeled "}A\text{,"} \\ (\sigma_9) & \text{A circle labeled "}D\text{" overlaps a circle labeled "}A\text{"} \end{array} \right\}$$

FIGURE 49 Projection of a disjunctive constraint underlying an over-specificity potential of Euler diagrams.

$\sigma_7\}$. As these source types collectively indicate the given premises, $\{\theta_9, \theta_{10}, \theta_{11}\}$, we can say that the resulting diagram expresses these assumptions. Now, because of the structural constraint holding on Euler diagrams (symbol $\vdash$ connecting $\{\sigma_5, \sigma_6, \sigma_7\}$ to $\{\sigma_8, \sigma_9\}$), our Euler diagram must support either σ_8 or σ_9. Since these respectively indicate θ_{12} and θ_{13}, our diagram inevitably expresses either information θ_{12} or information θ_{13}. Unfortunately, within standard set theory, neither information follows from premises $\{\theta_9, \theta_{10}, \theta_{11}\}$.

Moreover, on the semantic convention for Euler diagrams, $\{\sigma_5, \sigma_6, \sigma_7\}$ is the only set of source types that collectively indicate assumptions $\{\theta_9, \theta_{10}, \theta_{11}\}$. It follows that *no* Euler diagrams can express these assumptions without thereby presenting either θ_{12} or θ_{13}.

The Case of Euler Diagrams (Case 2)

The second case is a similar story. After drawing an Euler diagram (Figure 46) to express premises θ_9 and θ_{10}, we wish to add premise θ_{16} to the diagram. For this purpose, we need to realize the following source type in the diagram:

(σ_{10}) A circle labeled "*D*" encloses a circle labeled "*A*" in it,

while preserving source types σ_5 and σ_6. *As a matter of structural constraint* governing Euler diagrams, however, whenever an Euler diagram supports every source type in $\{\sigma_5, \sigma_6, \sigma_{10}\}$, it must support either of the following source types:

(σ_{11}) The circle labeled "*D*" has no overlap with circles labeled "*B*" and "*C*."

$$\left\{\begin{array}{ll}(\theta_9) & C \subset B,\\ (\theta_{10}) & B \cap A = \emptyset,\\ (\theta_{16}) & A \subset D\end{array}\right\} \qquad \left\{\begin{array}{ll}(\theta_{17}) & D \cap B = \emptyset \text{ and } D \cap C = \emptyset,\\ (\theta_{18}) & D \cap B \neq \emptyset \text{ and } D \cap C = \emptyset,\\ (\theta_{19}) & D \cap B \neq \emptyset \text{ and } D \cap C \neq \emptyset\end{array}\right\}$$

$$\Uparrow \qquad\qquad\qquad\qquad \Uparrow$$

{ (σ_5) A circle labeled "C" is inside a circle labeled "B," (σ_6) A circle labeled "A" is completely outside the "B" circle, (σ_{10}) A circle labeled "D" encloses the "A" circle in it } $\vdash$ { (σ_{11}) The circle labeled "D" has no overlap circles labeled "B" and "C," (σ_{12}) The circle labeled "D" overlaps a circle labeled "B" but not with a circle labeled "C," (σ_{13}) The circle labeled "D" overlaps circles labeled "B" and "C" }

FIGURE 50 Projection of a disjunctive constraint underlying another over-specificity potential of Euler diagrams.

(σ_{12}) The circle labeled "D" overlaps a circle labeled "B" but not a circle labeled "C."

(σ_{13}) The circle labeled "D" overlaps circles labeled "B" and "C."

On the semantic convention we adopt, these alternative source types respectively indicate information θ_{17}, θ_{18}, and θ_{19}. Unfortunately, none of these follows from the given premises $\{\theta_9, \theta_{10}, \theta_{16}\}$. Thus, we must either give up on expressing premise θ_{16} or tolerate our diagram to express one of the above invalid conclusions. Again, a structural constraint governing Euler diagrams gives rise to particular over-specificity of the system, namely, specificity about alternatives θ_{17}, θ_{18}, and θ_{19} in expressing information $\{\theta_9, \theta_{10}, \theta_{16}\}$. Figure 50 schematizes this situation.

The General Case

We will now describe, in general terms, how a structural constraint on representations gives rise to an over-specificity potential of a system. We start by explaining what a disjunctive constraint is.

Definition 3 (Disjunctive Constraint)
A *disjunctive constraint* is a constraint, $\Theta_1 \vdash \Theta_2$, such that:

1. Θ_2 has more than one members,
2. $\Theta_1 \nvdash \Theta_2'$ for any proper subset Θ_2' of Θ_2.

Remember that $\Theta_1 \vdash \Theta_2$ is defined as a constraint that if *all* members of Θ_1 holds, *some* member of Θ_2 must hold. Clause 1 then says that there are in fact more than one alternative in Θ_2. So, whenever all members of Θ_1 hold, some of the members of Θ_2 must hold, but the member or members that hold can vary case by case. Clause 2 then says that each alternative in Θ_2 is necessary to make this constraint hold—a reduced set of alternatives is not enough to cover all cases in which Θ_1 holds.

For an example of a disjunctive constraint, take the lower constraint shown in Figure 50. If σ_5, σ_6, and σ_{10} hold of an Euler diagram, either σ_{11}, σ_{12}, or σ_{13} must hold of it. Reduce the set of alternatives to σ_{12} and σ_{13}. Then, the constraint no longer holds, since even if σ_5, σ_6, and σ_{10} hold of an Euler diagram, it is quite possible that neither σ_{12} nor σ_{13} holds, and σ_{11} holds instead. The same applies however we reduce the set of alternatives into a smaller set. Each alternative σ_{11}, σ_{12}, and σ_{13} must be listed in the consequent for the constraint to hold. Thus, $\{\sigma_5, \sigma_6, \sigma_{10}\} \vdash \{\sigma_{11}, \sigma_{12}, \sigma_{13}\}$ is a disjunctive constraint.

We are now ready to characterize the semantic property of a representation system that we have informally referred to as a "projection of a disjunctive constraint."

Definition 4 (Projection of a Disjunctive Constraint) Let Γ_1, Γ_2 be sets of source types of a representation system $\mathcal{R}$. We say $\mathcal{R}$ *projects the disjunctive constraint* $\Gamma_1 \vdash \Gamma_2$ if:

1. $\Gamma_1 \vdash \Gamma_2$ is a disjunctive constraint,
2. There are sets Δ_1 and Δ_2 such that:
   (a) $\Gamma_1 \Rightarrow \Delta_1$,
   (b) $\Gamma_2 \Rightarrow \Delta_2$.

Figure 51 shows this general condition schematically. Under this condition, system $\mathcal{R}$ has a potential for over-specificity, i.e., the expression of Δ_1 with Γ_1 forces us to express an additional piece of information that may not follow from Δ_1. To wit, suppose we want to express Δ_1 in a representation of system $\mathcal{R}$. By clause 2 (a), realizing Γ_1 in our representation is a way of doing so. By clause 1 above, however, a structural constraint from Γ_1 to Γ_2 holds on the representation. Thus, we are forced to choose some member of Γ_2 to realize when we realize Γ_1. By clause 2 (b), whichever member of Γ_2 we choose, our representation ends up expressing a member of Δ_2 as an additional piece of information. Yet it may well be that no member Δ_2 is a consequence of Δ_1. Our examples in Section 3.1 have demonstrated that this is often the case.

A direct consequence of the projection of a disjunctive constraint is the danger of accidental features. Faced with the need to express the set of information Δ_1 in system $\mathcal{R}$, one might go ahead and realize the set of source types

$$\begin{array}{ccc} \Delta_1 & & \Delta_2 \\ \Uparrow & & \Uparrow \\ \Gamma_1 & \vdash & \Gamma_2 \end{array}$$

FIGURE 51 Projection of a disjunctive constraint in a representation system.

Γ_1 in one's representation. This of course results in a representation supporting one of the alternative source types in Γ_2. The chosen alternative, γ_i, as well as any source type entailed by the realization of Γ_1 plus γ_i, becomes an *accidental feature* of our representation. We have illustrated various such accidental features with examples of memory maps, geometry diagrams, and block drawings.

These features are dangerous because once realized, no visual features distinguish them from other properties that indicate the premises or valid conclusions obtained by free rides. Yet, once they are interpreted according to the semantic rule of the representation system, they lead one to invalid conclusions most of the time, except when they happen to indicate the consequences of Δ_1.

Now, when a disjunctive constraint $\Gamma_1 \vdash \Gamma_2$ is projected as in Definition 4, it is often the case that Γ_1 is the only set of source types that icollectively indicates Δ_1. This means that realizing all source types in Γ_1 is the only way to express all target types in Δ_1. In the first case of Euler diagrams, for example, $\{\sigma_5, \sigma_6\}$ is the only set of source types that collectively indicates $\{\theta_9, \theta_{10}\}$, so realizing σ_5 and σ_6 in a diagram is the only way to express information θ_9 and θ_{10} in this system of Euler diagrams.

Generally, when this "only-ness" condition holds for Γ_1 of a disjunctive constraint $\Gamma_1 \vdash \Gamma_2$ projected in a representation system, the system has *no* way of expressing all information in Δ_1 without expressing some unwarranted information. If there were some alternative set of source types, say Γ_1', that collectively indicates Δ_1, and if the conditions for Γ_1 in Definition 4 does not hold for this Γ_1', then the system could express all information in Δ_1 with this set Γ_1', without thereby expressing unwarranted information. Unfortunately, with the exception of the case of Hyperproof diagrams, all examples of over-specificity discussed in Section 3.1 are cases where the relevant system has no alternative set of source types. We say a representation system *strongly* projects a disjunctive constraint $\Gamma_1 \vdash \Gamma_2$ if all the conditions in Definition 4 are met while Γ_1 is the only set of source types in the system such that $\Gamma_1 \Rightarrow \Delta_1$.

This strong projection of a disjunctive constraint amounts to something that may be called a "limitation on selective expression." This is not the same as a lack of expressive coverage, which refers to inadequacy of the sheer range of information that can be expressed in the system. This range may be large enough, but some sets of information in the range cannot be expressed exclusively, without additional information also being expressed.

Levesque (1988) clarified that being flexible in this respect is an important asset of a representation system. His foremost example of such a flexible system was first-order languages. He stated:

> As far as knowledge representation is concerned, the representational expressiveness of a language like that of first-order logic is not so much in what it allows you to say, but in what it allows you to *leave unsaid*. (Levesque 1988: p. 370, emphasis by the original author.)

For example, the system of Euler diagrams does not allow us to leave something unsaid: when we are to express θ_9, θ_{10}, and θ_{11}, we must also express either θ_{12} or θ_{13}; when we are to express θ_9, θ_{10}, and θ_{16} in the system, we must also express either θ_{17}, θ_{18}, or θ_{19}. In contrast, we can express θ_9, θ_{10}, and θ_{11} in a first-order language with suitable predicates and names without expressing any additional information. In fact, the following text is such an expression in the first-order language of set theory:

$$C \subset B. \quad B \cap A = \emptyset. \quad B \subset D.$$

Or recall the example of position diagrams discussed in Chapter 1, where we witnessed English text could express certain pieces of information exclusively while position diagrams could not. Generally, a language (a linguistic representation system) has this strength of selective expression, over many diagrammatic representation systems.

3.2.2 Application of the Notion to Specific Examples

The Case of Memory Maps

This is the case in which the system of Harry's memory map exhibits over-specificity about alternatives θ_2, θ_3, and θ_4 in expressing θ_1 in the map in Figure 52.

If our analysis is correct, we should be able to specify a structural constraint governing Harry's maps that is responsible for this over-specificity. Consider the constraint from the set of source types:

(σ_{14}) The road line 1 is between the *A* and *L* blocks along the river band.

(σ_{15}) The *L* block is between road line 1 and the *K* block along the river band.

(σ_{16}) The *B* block is between the *A* and *K* blocks along the river band, without any overlap with road lines and other house blocks.

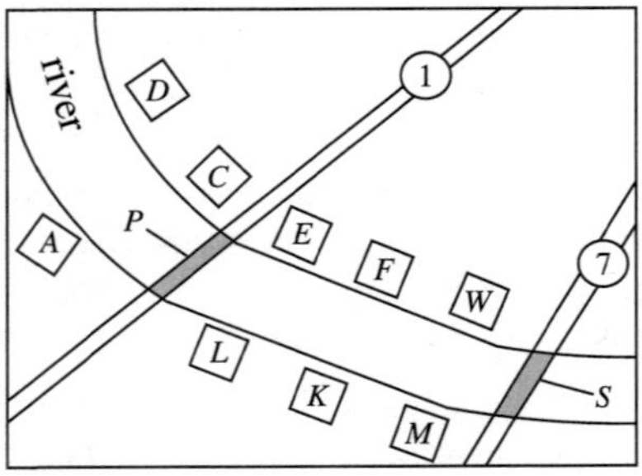

FIGURE 52 Harry's memory map that turns over-specific if information θ_1 is added. Shown again.

to the disjunction of source types:

(σ_1) The *B* block is between the *A* block and road line 1.

(σ_{18}) The *B* block is between road line 1 and the *L* block.

(σ_{19}) The *B* block is between the *L* and *K* blocks.

The reader should check that, assuming the *B* block does not come on top of the road line or the other blocks, Harry's memory maps is governed by this structural constraint. Of these source types, σ_{14} and σ_{15} already hold of the memory map in Figure 52.

Recall that we let Θ_m be the entire set of information expressed in this map. Let Σ_m be the set of source types that collectively indicates this set of information Θ_m. Then, σ_{14} and σ_{15} are members of Σ_m, and since the constraint $\{\sigma_{14}, \sigma_{15}, \sigma_{16}\} \vdash \{\sigma_1, \sigma_{18}, \sigma_{19}\}$ holds, it follows that the constraint from $\Sigma_m \cup \{\sigma_{16}\} \vdash \{\sigma_1, \sigma_{18}, \sigma_{19}\}$ also holds on Harry's maps. It is this constraint that is responsible for the over-specificity in question.

Harry wants to add the piece of information θ_1 by realizing the state of affairs σ_{16} in this map. However, due to this structural constraint on the map, if he realizes σ_{16} while preserving the source types Σ_m, he must realize one of the alternative states of affairs σ_1, σ_{18}, and σ_{19}. These source types respectively indicate the pieces of information θ_2, θ_3, and θ_4. Unfortunately, neither follows from the set of information $\Theta_m \cup \{\theta_1\}$ that Harry is willing to keep in his memory map. Moreover, according to the semantic convention associated with Harry's memory map, there is no other way of putting this information in a memory map. More precisely, there is no other set of source types than $\Sigma_m \cup \{\sigma_{16}\}$ that collectively indicates the set of information $\Theta_m \cup \{\theta_1\}$. Thus, the system of Harry's memory maps strongly projects the disjunctive constraint $\Sigma_m \cup \{\sigma_{16}\} \vdash \{\sigma_1, \sigma_{18}, \sigma_{19}\}$. It is over-specific in presenting the set of information $\Theta_m \cup \{\theta_1\}$, exactly in the way our analysis characterizes.

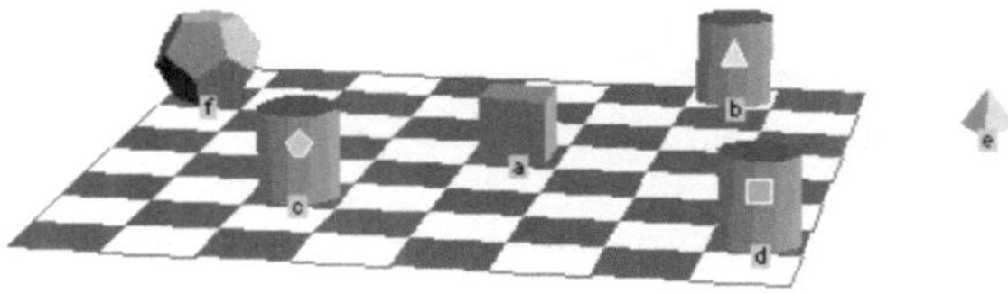

FIGURE 53 Hyperproof diagram that turns over-specific if the sentence `SameCol(e,a)` or `SameSize(b,f) ∨ SameSize(c,f)` is applied. Shown again.

The Case of Hyperproof Diagrams

We saw that the system of Hyperproof diagrams is over-specific in presenting the content of each of the following sentences in the diagram in Figure 53.

- `SameCol(e,a)`
- `SameSize(b,f) ∨ SameSize(c,f)`

Let us look at the case of the first sentence. Here is the information presented in the sentence:

(θ_{20}) The *e* block is in the same column as the *a* block.

Since we let Θ_h be the entire information content of the diagram, we are considering over-specificity in the system of Hyperproof diagrams in expressing the set of information $\Theta_h \cup \{\theta_{20}\}$. Consider the constraint, holding of Hyperproof diagrams, from the set of source types:

(σ_{20}) The cube icon labeled "*a*" is in row 5 and column 5.

(σ_{21}) The tetrahedron icon labeled "*e*" is in the same column as the cube icon labeled "*a*."

to the disjunction of source types:

(σ_{22}) The tetrahedron icon labeled "*e*" is in row 1 and column 5.

(σ_{23}) The tetrahedron icon labeled "*e*" is in row 2 and column 5.

(σ_{24}) The tetrahedron icon labeled "*e*" is in row 3 and column 5.

(σ_{25}) The tetrahedron icon labeled "*e*" is in row 4 and column 5.

(σ_{26}) The tetrahedron icon labeled "*e*" is in row 6 and column 5.

(σ_{27}) The tetrahedron icon labeled "*e*" is in row 7 and column 5.

(σ_{28}) The tetrahedron icon labeled "*e*" is in row 8 and column 5.

The reader should convince him/herself that this is a structural constraint that is reasonable to assume on Hyperproof diagrams. Here, σ_{20} is one of the source types that already hold of the diagram above.

Recall that we let Θ_h be the entire set of information expressed in the diagram in Figure 53. Let Σ_h be the set of source types, holding of the diagram, that indicate this chunk of information. Since σ_{20} is a member of Σ_h, the constraint from $\Sigma_h \cup \{\sigma_{21}\}$ to $\{\sigma_{22}, \sigma_{23}, \sigma_{24}, \sigma_{25}, \sigma_{26}, \sigma_{27}, \sigma_{28}\}$ holds on Hyperproof diagrams. Because of this constraint, when we add information θ_{20} by realizing source type σ_{21} in the above diagram, we must realize one of source types σ_{22}–σ_{28}. Each of these indicates a piece of information about the exact row in which tetrahedron *e* is located, but, unfortunately, none follow from the set of information $\Theta_h \cup \{\theta_{20}\}$ that we wish to express in the diagram. Furthermore, under the semantic rules for Hyperproof diagrams, there is no set of states of affairs, other than $\Sigma_h \cup \{\sigma_{21}\}$, that indicates the set of information $\Theta_h \cup \{\theta_{20}\}$. Thus, the system of Hyperproof diagrams strongly projects the disjunctive constraint $\Sigma_h \cup \{\sigma_{21}\} \vdash \{\sigma_{22}, \sigma_{23}, \sigma_{24}, \sigma_{25}, \sigma_{26}, \sigma_{27}, \sigma_{28}\}$. This is, therefore, a case of over-specificity as we have characterized in our analysis. A similar account explains why we cannot incorporate the content of the sentence `SameSize(b, f)` ∨ `SameSize(c, f)` into the given diagram without committing ourselves to some unwarranted information.

Reflecting this limitation in its diagrammatic part, the logical system underlying Hyperproof does not allow us to incorporate the contents of `SameCol(e,a)` or `SameSize(b,f)` ∨ `SameSize(c,f)` to the diagram. This is to prevent the diagram from supporting accidental features.

Thus, the diagrammatic part of the Hyperproof representation system is ineffective for the purpose of expressing an *arbitrary* combination of information we obtain (or assume) in a given stage of problem solving. However, the logical system of Hyperproof has a special rule, called "Case Exhaustive," that lets us consider several diagrams that embody the alternative source types, and draw a valid conclusion that is common in all these diagrams. Furthermore, the sentential part of the system supplements this limitation of the diagrammatic part. This is suggestive of the utility of heterogeneous systems of representations that comprise two or more kinds of representations.

The Case of Geometry Diagrams

The structural constraint responsible for over-specificity of geometry diagrams in our example is the one from:

(σ_{29}) There is a triangle icon whose sides are labeled "*a*," "*b*," and "*c*."

to the disjunction of source types:

(σ_{30}) The "*a*" side of the triangle icon is longer than the "*b*" side.

(σ_{31}) The "*a*" side of the triangle icon is as long as the "*b*" side.

(σ_{32}) The "*a*" side of the triangle icon is shorter than the "*b*" side.

This disjunctive constraint is then projected through the indication relation

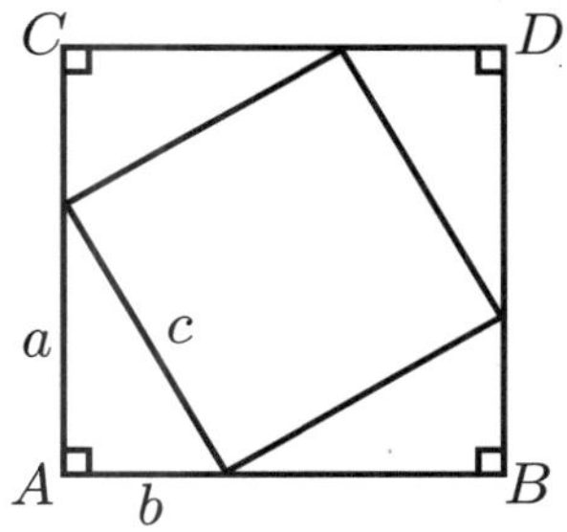

FIGURE 54 Geometry diagram used in a typical proof of the Pythagorean Theorem. Shown again.

associated with this system of geometry diagrams. Specifically, σ_{29}, σ_{30}, σ_{31}, and σ_{32} respectively indicate σ_{21}, σ_{22}, σ_{23}, and σ_{24}:

(θ_{21}) There is a triangle with sides *a*, *b*, and *c*.

(θ_{22}) Side *a* of the triangle is longer than side *b*.

(θ_{23}) Side *a* of the triangle is congruent with side *b*.

(θ_{24}) Side *a* of the triangle is shorter than side *b*.

Figure 55 shows the projection of the disjunctive constraint responsible for the present case of over-specificity. It clarifies that accidental feature in geometry diagrams (σ_{30}, σ_{31}, and σ_{32} in the present case) have their root in their over-specificity potentials, which in turn have their root in the structural constraint governing the geometry diagrams.

3.3 Questions and Responses

Let us consider some of the important questions that naturally arises concerning our analysis.

3.3.1 Question: Isn't It an Artifact?

You claim that accidental features of diagrams are invariably dangerous. However, this depends on semantic rules that you, as a theorist, have assumed for a representation system. Take the case of the geometry diagram discussed in Subsection 3.1.3. One of the accidental features of this diagram is that the line segment labeled "*a*" is longer than that labeled "*b*" (See Figure 45). However, this feature is dangerous only because you, as a theorist, have assumed a rather strong semantics for the system of geometry diagrams that interprets the relative length of line segments in the diagram. If, on the other hand, another theorist assumes weaker semantics, say one that ignores every quantitative feature of diagrammatic elements, the relative length of these line segments poses no risk of an invalid conclusion. In fact, under such

{ (θ_{21}) There is a triangle with sides *a*, *b*, and *c* }

$\Uparrow$

{ (σ_{29}) There is a triangle icon whose sides are labeled "*a*," "*b*," and "*c*" }

$\vdash$

{ (θ_{22}) Side *a* of the triangle is longer than side *b*,
(θ_{23}) Side *a* of the triangle icon is congruent with side *b*,
(θ_{24}) Side *a* of the triangle is shorter than side *b* }

$\Uparrow$

{ (σ_{30}) The "*a*" side of the triangle icon is longer than the "*b*" side,
(σ_{31}) The "*a*" side of the triangle icon is as long as the "*b*" side,
(σ_{32}) The "*a*" side of the triangle icon is shorter than the "*b*" side }

FIGURE 55 Projection of a disjunctive constraint underlying an over-specificity potential of geometry diagrams.

semantics, the example does not even constitute a case of over-specificity, since members of $\{\sigma_{30}, \sigma_{31}, \sigma_{32}\}$ no longer collectively indicate members of $\{\theta_{22}, \theta_{23}, \theta_{24}\}$ (the right $\Uparrow$ changes to $\not\Uparrow$ in Figure 55) and the disjunctive constraint $\{\sigma_{29}\} \Rightarrow \{\sigma_{30}, \sigma_{31}, \sigma_{32}\}$ is not projected.

Or take a typical pictorial illustration in a popular science magazine that explains the mechanism behind a natural phenomenon (Figure 56). It is designed to show a common mechanism behind different occurrences of the phenomenon, but doing so in the form of pictorial illustrations requires the illustrator to "fill in the details" that are not part of the common mechanism, such as the shapes, colors, and positions of clouds. According to your analysis, the graphical features depicting these fill-ins are accidental features of the illustration, and are supposed to be dangerous. However, few regular readers of science magazines take these features seriously to indicate something general about the phenomenon. They appear to carry meaning only because a theorist has assumed rather unrealistic semantics that interprets the shapes, colors, and positions of the cloud pictures and other inessential details. In general, whether accidental features pose real dangers, or whether a representation system is really over-specific, depends on what semantic conventions a theorist associates with the system. Some, perhaps all, cases of apparent over-specificity are theorists' artifacts in this sense.

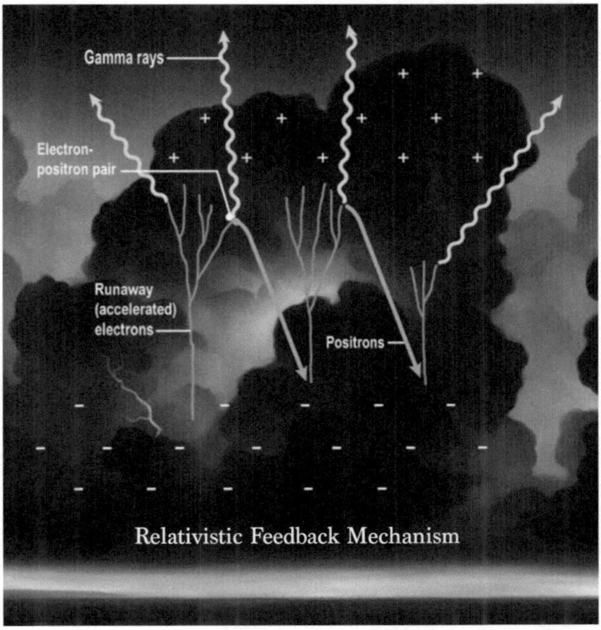

FIGURE 56 Typical pictorial illustration in popular science magazines. Reprinted from Dwyer and Smith (2012) with permission of Brian Despain. The original illustration has been trimmed and made into black and white in republication.

Response

We agree that whether accidental features pose real dangers and whether a system is really over-specific depends on the semantic conventions of the system, but do not agree that a theorist can assume arbitrary semantic rules to make up a case of accidental features or over-specificity. A representation system is a set of regularities with which a group of people is engaged in a representational practice, and what semantic conventions are adhered to in this practice is an objective matter that a theorist needs to investigate, rather than just assume.

From this point of view, what the case of geometry diagrams demonstrates is that people have developed a modified semantic convention that leaves certain quantitative features of the geometry diagrams uninterpreted. We typically acquire this convention through training in high-school geometry, and learn to read and draw geometry diagrams on the assumption that every learned person has complied with this convention. This gives rise to a new system of geometry diagrams, and under its modified semantic rules, geometry diagrams have neither over-specificity potentials nor accidental features.

Giaquinto (2007) forcefully argues that under such a sophisticated interpretation, an argument in geometry that involves a diagram can be a real proof. In fact, Luengo (1995) started an investigation into a logical system that formalized the rules for valid uses of geometry diagrams. This project was later taken up by Miller (2007) and Mumma (2010), contributing to our understanding of the exact roles of diagrams in geometry proofs under the sophisticated interpretation. Nevertheless, the fact remains that there exists a primitive system of geometry diagrams with more literal semantic rules, and that minimization of the over-specificity problems in *this* system has driven the development of a modified system. Neither the primitive system nor its over-specific character is an artifact based on a theorist's arbitrary assumption.

The case of pictorial illustration is similar. It clearly shows that the readers and the illustrators have developed a semantic convention that leaves certain details of the illustration uninterpreted, and that a more literal convention for illustrations serves as the base from which the alternative semantics has developed. The alternative semantics is an upper layer of this base semantics, so to speak, just as the modified semantics of geometry diagrams is an upper layer on the more primitive semantics.

In the case of pictorial illustrations in magazines, however, the modified convention is less standardized than that for geometry diagrams, and what exact features of illustrations belong to the upper layer are not always obvious to the reader. Thus, a more realistic view of this case is that people have developed a subsidiary system of signs or cues that define this "uplifted" semantics *in situ*. Such a system utilizes contextual cues such as text annotating an illustration, our common knowledge, and so-called "graphical implicatures" (Marks and Reiter 1990; Oberlander 1996). The topic of how such an "uplifted" semantics develops from base semantics, as well as how it is supported by a subsidiary cuing system, is beyond the scope of this book. Nevertheless, it seems clear that people often adopt alternative semantic conventions on syntactically identical sets of representations. The adoption of such systems can be quite dynamic, but it should not be mistaken for the situation where a theorist assumes arbitrary sets of semantic rules case by case.

3.3.2 Question: Are Free Rides and Over-Specificity So Different?

This is a continuation of the objection raised in Subsection 2.4.3. You showed that an over-specificity potential limits a system's ability of selective expression of information, but for that matter, a free ride potential limits selective expression of information. True, the additional piece of information that is expressed in a free ride is a consequence of the given set of information, but it still prevents one from expressing a particular set of information exclusively, without thereby expressing some other information. In this respect, free rides and over-specificity are on a par: they both reduce the user's discretion over

$$
\begin{array}{ccccccc}
\Delta & \vdash & \delta_i & \Delta_1 & & \Delta_2 \\
\Uparrow & & \Uparrow & \Uparrow & & \Uparrow \\
\Gamma & \vdash & \gamma_i & \Gamma_1 & \vdash & \Gamma_2 \\
 & \text{(a)} & & & \text{(b)} &
\end{array}
$$

FIGURE 57 Semantic property underlying (a) a free-ride potential and (b) an over-specificity potential.

the information expressed in his or her diagram, and are counter to the ideal of controlling the information involved in his or her inference. They are not as different as your analysis suggests.

Response

They are different potentials, but our analysis also revealed exactly what is common between them. Let us compare the two semantic properties underlying these phenomena (Figure 57-a for the free ride potential and Figure 57-b for the over-specificity potential). Both figures shows a structural constraint on representations at the bottom, and both the antecedent and consequent of this constraint have arrows pointing up to (sets of) target types. This common structure, which we call a *constraint projection*, captures the mechanism by which structural constraints on representations intervene in the system's ability of selective expression.

Specifically, the constraint projection in Figure 57-a makes the expression of Δ_1 through Γ_1 entail the expression of δ_i through γ_i; the constraint projection in Figure 57-b makes the expression of Δ_1 through Γ_1 entail the expression of a member of Δ_2 through a member of Γ_i. In other words, our analysis explains, not annihilates, the limitations of selective expression commonly entailed by free ride potentials and over-specificity potentials.

In Subsection 2.4.3, Frege (1879/1967) was cited to show that one ideal for the use of external representations in inference is to *control* the information involved in our inference, as opposed to *making inference efficient*. Our analysis revealed that representation systems with a constraint projection property limit selective expression and hence invariably interferes with the ideal of control. This much is common to free ride potentials and over-specificity potentials. When a constraint projection takes the form of a consequence tracking, however, it serves the ideal of efficiency by providing free rides in inference. When a constraint projection takes the form of projection of a disjunctive constraint, it interferes with the ideal for efficiency by taxing the user in choosing what to express and, in some cases, forcing him or her to put accidental features into his or her representation.

3.4 Chapter Takeaways

- Over-specificity is the potential of a representation system, where expressing a set of information in diagrams mandates the selective expression of other, often non-consequential pieces of information.
- Over-specificity of a representation forces us to either refrain from expressing the set of information in diagrams or tolerate the presence of an additional piece of unwarranted information in our diagram.
- Many diagrammatic representation systems have over-specificity, including systems of position diagrams, Euler diagrams, memory maps, block drawings, geometry diagrams, and pictorial illustrations.
- Over-specificity holds in a given representation system when, due to constraints on spatial configurations of representing objects, the unique set of source types that collectively indicates a set of target types entails a disjunction of source types, with each disjunct indicating a target type that may not follow from the original set of target types.
- So-called "accidental features" of a diagram are those source types that come to hold of the diagram as a result of expressing a set of information as to which the system of diagrams is over-specific.

4

Potential for Consistency-Check

> It is not possible to express a certain range of inconsistent sets of information in diagrams. Thus, drawing a diagram can serve as the basic consistency check of a set of information that the diagram expresses, substituting for the mental process of evaluating the consistency of the set in question.

This chapter investigates the third main theme outlined in Chapter 1, which has been frequently cited or alluded to in modern discussions of cognitive potentials of diagrams. We start with the description of the phenomenon and move on to an explication of the semantic properties of diagrams that underly the phenomenon.

4.1 The Phenomenon

4.1.1 Schön (1983): Cross-Section Diagrams

This example is concerned with the same architectural-design situation discussed back in Chapter 2 (Section 2.1.5), where an architect, Quist, was helping an architectural student, Petra, with the design of elementary-school buildings (Schön 1983). Petra's basic idea was to construct a class-room building with three consecutive L-shaped components (Figure 58) but she had difficulty in figuring out how to fit the building to the slopes of the land. She tried to "butt the shape of the building into contours of the land" (p. 82), but this solution did not seem to work.

While retaining her basic idea, Quist suggested her a different way of framing the problem of fitting—from how to butt the bottom of the building into the existing shape of the slope to how to impose the parallel structure of the building's roof and bottom into the slope. Quist proceeded to justify his suggestion by showing Petra that it is possible to impose the parallel structure to the slope of the land. Schön (1983) perceives it as a "hypothesis-testing experiment," where "the hypothesis is that this slope and this geometry of parallels can be made to fit each other" (p. 149).

Note that the inferential goal at this stage is to prove a *possibility* (that this

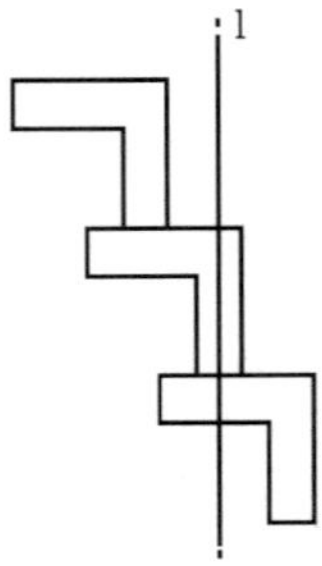

FIGURE 58 Petra's basic idea of a class-room building with three consecutive L-shaped components. The dotted line shows the sectioning line used for Figure 59.

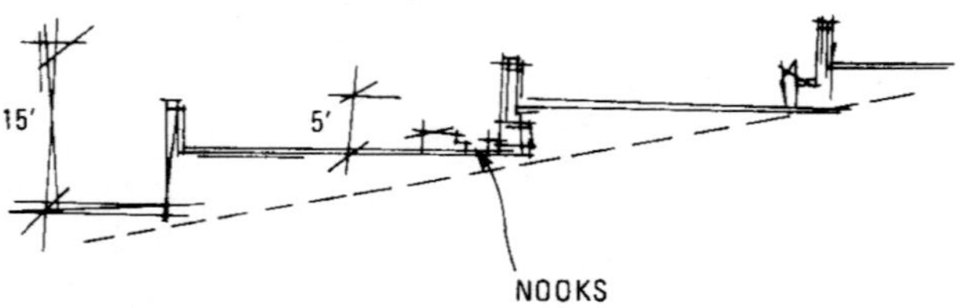

FIGURE 59 Cross-section diagram proving the possibility that two consecutive lands be fitted to the slope, each with sufficient depth for an L-shape classroom Petra had in mind. The direction of the cross section is indicated by line *l* in Figure 58. The right means the north. Reprinted with permission of Perseus Books Group, from Schön (1983); permission conveyed through Copyright Clearance Center, Inc.

slope and this geometry of parallels *can* be made to fit each other). It is in contrast to the stage discussed in Chapter 2, where Quist finds a *consequence* (that the roof of the second-level building rise considerably from the base of the third level) of his design ideas.

Toward this goal, Quist drew a cross-section diagram that showed four consecutive flat terraces fitted to the slope line, two of which were shown to be deep enough for an L-shape building that Petra had in mind. Figure 59 shows the cross-section diagram that Quist drew, where the right means the north.

To be more specific about the informational content of this diagram, let *A*, *B*, *C*, and *D* be the four consecutive terraces drawn in the diagram in the order from right to left. The diagram shows the following pieces of information among others:[15]

[15] The diagram was meant to express only approximate directions, lengths, and degrees, so all specifications of these variables in the following discussion should be taken as such.

(θ_1) Terrace *B* is horizontal and 24-foot deep.

(θ_2) Terrace *C* is horizontal and 24-foot deep.

(θ_3) The north end of *B* is connected to the south end of *A* with a vertical wall.

(θ_4) The north ends of terraces *B*, *C*, and *D* make a $\frac{5}{24}$ slope with its south end down.

Clearly, both Quist and Petra took this cross section diagram as proof that such a set up of terraces along the slope was possible, and this information eventually led them to the conclusion that the L-shape classrooms can be fitted to the slope. Quist's hypothesis had thus been confirmed.

Notice, however, that the cross-section in Figure 59 was just a *representation* of the land on a piece of paper, not the real land. If one actually sets up actual terraces on the real land, then it would trivially imply that such a set-up of terraces is possible on the land. ("Truth implies possibility.") In the present case, a mere representation of such a set-up works as proof of its possibility. This is not a trivial matter.

This capacity of cross-section diagrams comes from the fact that they cannot express geometrically impossible set-ups of terraces and other physical objects. For example, change condition θ_3 to the following slightly more specific condition:

(θ_5) The north end of *B* is connected to the south end of *A* with a vertical wall of 3 feet.

The new set of conditions, $\{\theta_1, \theta_2, \theta_4, \theta_5\}$ with θ_5 instead of θ_3, is a geometrically impossible set-up of terraces on the land. (The vertical raise of the south end of *B* must be approximately 5 feet.) *Correspondingly*, the present system of cross-section diagrams cannot express it. We can easily think of other examples of geometrically impossible set-ups of lands that cannot be expressed by using cross-section diagrams. Because of this general limitation of cross-section diagrams, once we find a cross-section diagram expressing an arrangement of objects, it proves that the arrangement is geometrically possible, at least at the sections represented by the diagram.

We witness the consistency check potential of diagrams in non-professional design situations too. A client may sketch a floor map to check the *possibility* of placing a bathroom just inside the house entrance before he or she proposes it to the architect. A wife may draw a sketch such as that in Figure 60 to prove to her husband the *possibility* of creating a direct pathway across the living room (Shimojima 2002). Furniture shops on the web take advantage of exactly this consistency check potential of diagrams when they offer a diagram editor (variously called "room planner," "furniture simulator," etc.) where the potential buyer can confirm that a furniture item can fit into a room.

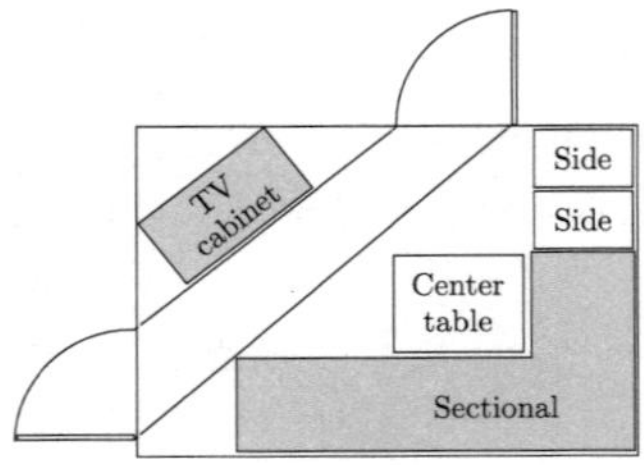

FIGURE 60 Furniture arrangement map that demonstrates the possibility of creating a direct pathway across a living room (Shimojima 2002).

A furniture map cannot express a geometrically impossible two-dimensional arrangement of furniture, and that is why such a map works as proof of the possibility of the particular arrangement that it expresses.

Generally, if a representation system cannot express a certain range of inconsistent sets of information, it is said to be *auto-consistent* with respect to that range of inconsistencies.[16] Here, it is important to note that a system of diagrams is auto-consistent relative to some particular range of inconsistent sets of information, but there always can be some other range of inconsistent sets of information relative to which the system is not auto-consistent. Relative to such a range of inconsistencies, the system can express an inconsistent set of information, and so the existence of a digram is no guarantee that the set of information it expresses is consistent. For example, the system of furniture maps guarantees only geometrical possibilities of two-dimensional arrangements. They do not guarantee that an arrangement is *three-dimensionally possible* (e.g., the ceiling in a certain part of a room may be too low for a bookshelf) nor that it is *psychologically* or *culturally* possible beside being geometrically possible (e.g., having the sectional conglomerated with two side tables as in Figure 60 may not be acceptable to the couple).

Returning to Schön's example, this limitation also applies to the system of cross-section diagrams that Quist drew. It only guarantees geometrical possibility of a two-dimensional set-up of consecutive terraces along the slope. It does not guarantee that the set-up is possible three-dimensionally nor with respect to other factors such as soil conditions and environmental protection. Consistency with such constraints is just not something the existence of a cross-sectional diagram can prove or disprove. Schön perceives this limitation of diagrams in the following way:

> Drawing functions as a context for experiment precisely because it enables the

[16]The term was first introduced by Barwise and Etchemendy (1994) when they discussed the system of Hyperproof diagrams. The auto-consistency of Hyperproof diagrams will be discussed in the next subsection.

> designer to eliminate features of the real-world situation which might confound or disrupt his experiments, but when he comes to interpret the results of his experiments, he must remember the factors that have been eliminated. (Schön 1983: p. 158)

The consistency check potential of diagrams is limited, but one can learn to take advantage of this potential despite its limitation. One can even take advantage of this limitation itself as a shield against confounding real-world factors.

4.1.2 Barwise and Etchemendy (1994): Hyperproof Diagrams

Generally, a set Δ of conditions is said to be consistent if it is possible that all members of Δ hold of a single object or situation. So, if $\Delta_1 \cup \{\neg\delta_i\}$ is consistent (where $\neg\delta_i$ is the negation of δ_i), it means that all members of $\Delta_1 \cup \{\neg\delta_i\}$ *can* hold of a single object or situation. This is equivalent to that all members of Δ_1 can hold of an object or situation without δ_i holding of it. In other words, the consistency of $\Delta_1 \cup \{\neg\delta_i\}$ means that δ_i is *not* a consequence of Δ_1.[17]

Thus, when a system $\mathcal{R}$ of diagrams cannot express a certain range of inconsistent sets of information, we can use this system to prove not only the consistency of a set of information but also the *non-consequence* of a piece of information *from* a set of information. Let $\perp_{\mathcal{R}}$ be the range of inconsistent sets of information that the system $\mathcal{R}$ cannot express, and suppose we find a diagram in $\mathcal{R}$ that expresses a set Δ_1 of information along with the negation of a piece of information δ_i. This means that $\Delta_1 \cup \{\neg\delta_i\}$ is not in the range $\perp_{\mathcal{R}}$ of inconsistencies. In other words, so as far as $\perp_{\mathcal{R}}$ is concerned, it is possible that δ_i does not hold of a situation where all members of Δ_1 hold. In this precise (but limited) sense, δ_i is a non-consequence of Δ_1.

The Hyperproof program (Barwise and Etchemendy 1994) takes advantage of this potential of diagrams to allow its user to prove non-consequence results, in addition to the usual consequence results. Recall that the diagrams coming with the Hyperproof program can depict blocks of varying sizes and shapes on an 8×8 chessboard. Suppose you are asked to show that θ_{13} is not a consequence of the premises θ_9–θ_{12}.

[17] For example, that Mantle is better than Mays and that Moskowitz is better than Mays can hold of a single situation while Moskowitz is not better than Mays in that situation. On the one hand, we express this fact by saying that $\{\theta_6, \theta_7\} \cup \{\neg\theta_8\}$ is consistent, where θ_6, θ_7, and θ_8 are the following pieces of information:

(θ_6) Mantle is better than Mays.

(θ_7) Moskowitz is better than Mays.

(θ_8) Moskowitz is better than Mays.

On the other hand, we can also express the above fact by saying that that Moskowitz is better than Mays does not follow from the information that Mantle is better than Mays and that Moskowitz is better than Mays. That is to say, θ_8 is not a consequence of $\{\theta_6, \theta_7\}$. Thus, the consistency of $\{\theta_6, \theta_7\} \cup \{\neg\theta_8\}$ means that θ_8 is not a consequence of $\{\theta_6, \theta_7\}$.

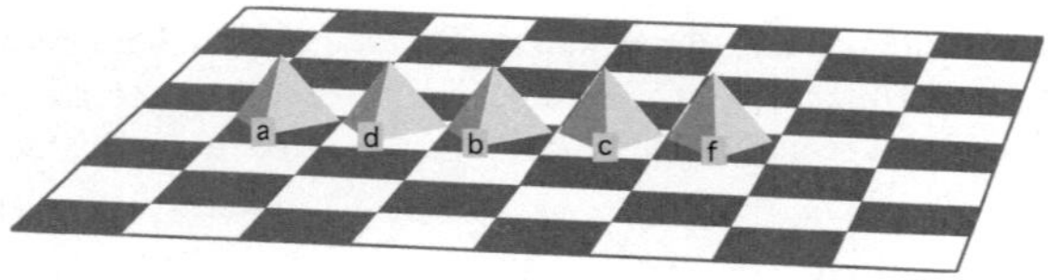

FIGURE 61 Hyperproof diagram that demonstrates the non-consequence of information θ_{13} from premises θ_9–θ_{12}.

(θ_9) Blocks *b* and *c* are in the same row.

(θ_{10}) Blocks *a* and *d* are in the same row.

(θ_{11}) Blocks *d* and *f* are in the same row.

(θ_{12}) Block *a* is to the left of block *b*.

(θ_{13}) Block *f* is to the left of block *c*.

If you have already started hand-drawing a diagram to find a "counter-example" to this argument, you are putting your diagram in exactly the same use Barwise and Etchemendy (1994) put their Hyperproof diagrams. Their program provides a diagram editor with which the user can draw a counter-example—a diagram that expresses all the premises θ_9–θ_{12} along with the negation of the purported conclusion θ_{13}, and when a user can draw such a diagram, the program accepts it as proof of the non-consequence of θ_{13} from θ_9–θ_{12}.

Figure 61 is an example of such a counter-example diagram. This diagram works as proof of non-consequence precisely because no Hyperproof diagrams can express a certain range of inconsistencies, that is, those sets of conditions that violate the constraints on the blocks world. Thus, by the logic we explained above, the existence of a diagram that expresses θ_9–θ_{12} along with the negation of θ_{13} guarantees that θ_{13} is not a consequence of θ_9–θ_{12}, as far as the constraints on the blocks world are concerned.

Compare this case with the case where the purported conclusion is in fact a consequence of the premises. One of such valid conclusions is the following:

(θ_{14}) Blocks *a* and *f* are in the same row.

Since θ_{14} is a consequence of θ_9–θ_{12}, it is impossible that θ_9–θ_{12} hold together with the *negation* of θ_{14}. And indeed, the system of Hyperproof diagrams cannot express θ_9–θ_{12} together with the negation of θ_{14}. If your diagram has icons labeled "*a*" and "*d*" in the same row while having an icon labeled "*f*" in the same row with an "*d*" icon, it is impossible not to have the "*a*" icon and the "*f*" icon in the same row. Thus, no Hyperproof diagram works as a counter-example to the consequence of θ_{14} from θ_9–θ_{12}. This is as expected since θ_{14} is indeed a consequence of θ_9–θ_{12}.

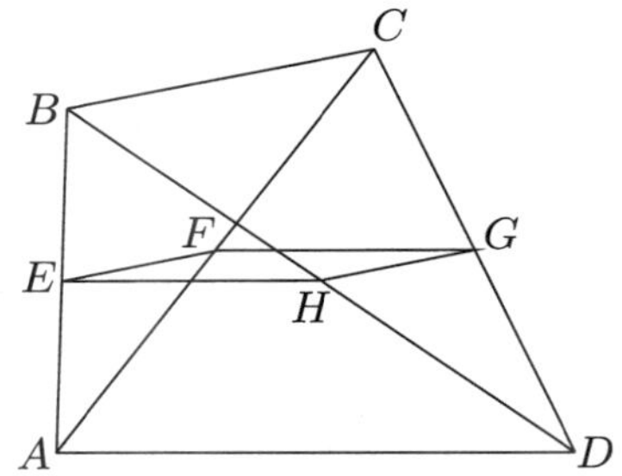

FIGURE 62 Geometry diagram that can be used in the Geometry Machine for consistency check.

Barwise and Etchemendy (1995a) gave formal semantics to the system of Hyperproof diagrams, and proved that it is auto-consistent with respect to the inconsistent sets of conditions violating the constraints of the blocks world. They pointed out that the language-based logical systems developed and studied by logicians typically focus on the proof of consequence as opposed to that of non-consequence. Establishing non-consequence results, however, is as important as establishing consequence results in actual practices of inference. Barwise and Etchemendy (1994) illustrate the point in the following way:

> When a defense attorney shows that the crime might have been committed by someone other than the client, say by the butler, the attorney is trying to prove that the client's guilt *does not* follow from the evidence in case. When mathematicians show that the parallel postulate is not a consequence of the other axioms of Euclidean geometry, they are doing the same thing: they are showing that it would be possible for the claim in question (the parallel postulate) not to be true, even if the other information (the remaining axioms) *is* true. (Barwise and Etchemendy 1994: p. 20, emphasis by the original authors.)

Curious as it may sound, diagrams can facilitate this important class of inferences by their *incapacity* of expressing a certain range of information.

4.1.3 Gelernter (1959): Geometry Diagrams

Gelernter (1959) used diagrams' potential of non-consequence proof to facilitate proof searches in his program for automated proof in geometry. The program, named "Geometry Machine," engages itself with the following problem, for example (Gelernter et al. 1960).

> Assume, in Figure 62, $ABCD$ is a quad-lateral, E the midpoint of $\overline{AB}$, F the midpoint of $\overline{AC}$, G the midpoint of $\overline{CD}$, and H the midpoint of $\overline{BD}$. Prove $EFGH$ is a parallelogram.

As many of us do, the Geometry Machine works backward: it starts from the goal statement ("$EFGH$ is a parallelogram") and looks for a set of subgoal statements that, if proved successfully, would be sufficient to establish the goal statement.

Typically, however, there are many such sets of subgoals for a given goal statement: the congruence of $\angle FEH$ and $\angle HGF$, the congruence of $\angle EFG$ and $\angle GHE$, the congruence of $\overline{FG}$ and $\overline{HE}$ along with their parallelism, the congruence of $\overline{FE}$ and $\overline{HG}$ along with their parallelism, and so on. Each of these subgoals then comes with different sets of sub-subgoals, which come with different sets of sub-sub-subgoals, and so on. Thus, for a given proof problem, there are many candidate proof sequences of subgoals that could yield the goal statement.

So, it is important for the program to have heuristics with which it can filter out unprovable subgoals (or sub-subgoals, etc.) before it looks for their proofs in vain. The Geometry Machine makes a crucial use of diagrams for this purpose. It "examines" the problem diagram (such as Figure 62) that expresses the premises and sees if the candidate sub-goal is "valid" in that diagram (Gelernter 1959: p. 142). If it is not valid, the program rejects the sequence including the sub-goal and starts working on another candidate sequence.

In our terminology, the case a subgoal is "not valid" in a diagram corresponds to the case where the *negation* of the sub-goal is expressed in the diagram. For example, suppose the program considers, as candidate subgoals, the congruence of $\overline{FE}$ and $\overline{HG}$ along with their parallelism. For the proof of the congruence of $\overline{FE}$ and $\overline{HG}$, the program considers proving the congruence of $\triangle FEA$ and $\triangle HGD$ by establishing $\overline{EA} \simeq \overline{GD}$, $\overline{AF} \simeq \overline{DH}$, and $\angle EAF = \angle GDH$. The diagram that comes with the problem, however, shows that $\overline{EA} \not\simeq \overline{GD}$ among others. Since the diagram expresses the premises of the problem, Gelernter (1959) takes it as an indication that the premises are compatible with the negation of the subgoal $\overline{EA} \simeq \overline{GD}$, and hence that $\overline{EA} \simeq \overline{GD}$ is not a consequence of the given premises. So, the program rejects the subgoal immediately, without making any attempt to find its proof.

Here, the crucial assumption is that when a geometry diagram expresses the set of premises along with the negation of $\overline{EA} \simeq \overline{GD}$, it indicates that the premises are compatible with the negation of $\overline{EA} \simeq \overline{GD}$. More generally, co-expressibility guarantees compatibility, and whatever set of information expressible together in a single geometry diagram is consistent. The system of geometry diagrams is auto-consistent with respect to the inconsistencies in the Eucledean space.

Thus, filtering of a subgoal in the Geometry Machine is really a non-consequence proof based on the consistency check potential of geometry diagrams. Under this particular application, non-consequence implies non-provability, so the consistency of a set of information recommends the rejection of a proposition as a target of proof.

Since the consistency check made by the Geometry Machine depends on the particular diagram that happens to come with the problem statement, it does not filter out every non-provable subgoal. For example, $\overline{EA} \simeq \overline{GD}$, rather than its negation, could be expressed in a diagram that comes with the problem. This, of course, is an *accidental feature* of a diagram discussed in Chapter 3, and in such a case, the non-consequential information $\overline{EA} \simeq \overline{GD}$ is accepted as a candidate subgoal.

Even so, the consistency check based on diagrams lets the program filter out many non-provable subgoals, saving a great deal of time and resources for the program. According to the estimate provided by Gelernter et al. (1960), the number of accepted subgoals at a given stage is reduced to $\frac{1}{200}$ on average, and so, "it is easy to see that the use of a diagram is crucial for our system" (p. 155).

4.1.4 Barwise and Etchemendy (1995): Euler Diagrams

In their unpublished, yet important paper, Barwise and Etchemendy (1995b) discusses an example of non-consequence proof with a Euler diagram. They consider the following problem cited from a psychological experiment on syllogistic reasonig (Johnson-Laird and Bara 1984):

> Some artists, bee-keepers, and chemists, at least one of each, are gathered in a room. None of the artists is a bee-keeper. All of the bee-keepers are chemists. Write down what, if anything, follows necessarily from these.

The experiment showed that 65 percent of the participants (13 out of 20 participants) mistakenly thought that these premises imply that no artist is a chemist. This result indicates that judging non-conseqeunce can be a difficult task even with these simple categorical propositions.

Drawing appropriate Euler diagrams, however, can help this judgement significantly. Figure 63 shows one of such Euler diagrams, redrawn from Barwise and Etchemendy (1995b) . It expresses all the premises θ_{15}–θ_{17} in the problem:

(θ_{15}) There are at least one artist, chemist, and bee-keeper.

(θ_{16}) No artists are bee-keepers.

(θ_{17}) All bee-keepers are chemists.

Yet, the circle labeled "artists" is not separated from the circle labeled "chemists," indicating the negation of the purported conclusion:

(θ_{18}) No artist is a chemist.

This Euler diagram therefore works as a counter-example to the purported conclusion.

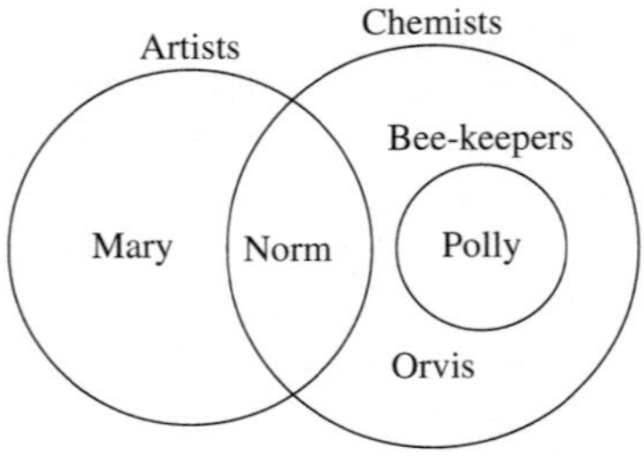

FIGURE 63 Euler diagram that demonstrates the non-consequence of information θ_{18} from premises θ_{15}–θ_{17}.

As with the previous examples of non-consequence proofs, this function of Euler diagrams comes from the fact that the system of Euler diagrams is auto-consistent with respect to the set-theoretic inconsistencies. Because of this general expressive limitation, the set of information expressed in the Euler diagram in Figure 63 is destined to be set-theoretically consistent. This in turn indicates that a piece of information is not a consequence of the given premises—not at least in the set-theoretic sense. Lemon and Pratt (1997) give a simple proof of the auto-consistency of a general-purpose system of Euler diagrams that expresses the distribution of objects to sets in general.

4.1.5 Further Examples: Bar Charts

The bar chart in Figure 64 shows the annual retention and graduation rates at four-year degree granting institutions in the US. The data are based on Aud et al. (2012). In this chart, the institutions are classified by their admission policies into five groups: *HS institutions* that accept less than 25% of applicants, *VS institutions* that accept 25.0–49.9%, *S institutions* that accept 50.0–74.9% applicants, *LS institutions* that accept 25% or more of applicants, and *OP institutions* that have no application criteria.

You can read off different levels of information from this bar chart. For example, the height of an individual bar lets you read off an approximate retention or graduation rate of a specific class of students for a specific group of institutions; the repeated pattern of downward shortening of each group of bars let us read off the general trend that groups of institutions with stricter admission policies have better rates in overall graduation, part-time student retention, and full-time student retention.

The fact that the bar chart lets us read off different levels of information is important and will be treated in the next chapter, but it is not our direct concern here. Our concern is the fact that, as long as the chart is correctly constructed and read, any set of information that we read off from it is guaranteed to be consistent at a certain basic level.

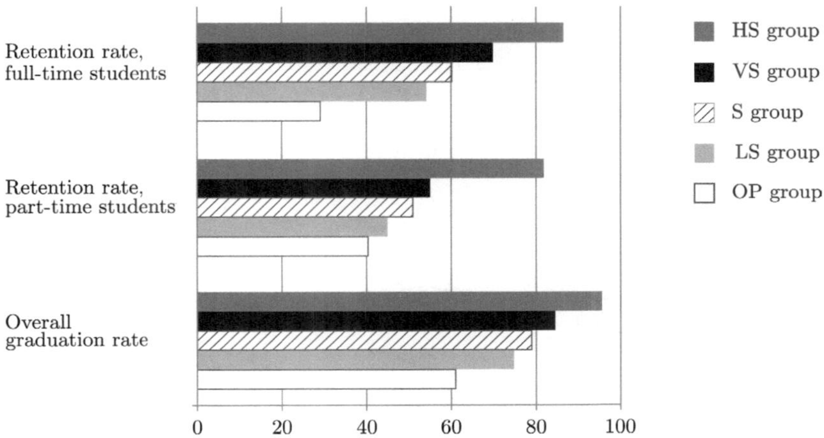

FIGURE 64 Bar chart showing the annual retention and graduation rates within six years at four-year degree granting institutions in the US. The data are based on Aud et al. (2012).

Underlying this guarantee is the general fact that a bar chart in this system cannot express a certain range of elementary inconsistencies. When it expresses that the overall graduation rate of HS institutions is around 87%, it never express that the value is around 67% (value-inconsistency); when it expresses that the retention of part-time students is higher in LS institutions than in OP institutions, it never expresses its opposite (violation of asymmetry); when it expresses that the retention of full-time students is higher in the VS institutions than the S institutions and that the retention of full-time students is higher in S institutions than LS institutions, it never expresses that the retention of full-time students is higher in LS institutions than in VS institutions (violation of transitivity or asymmetry). The system of bar charts is auto-consistent over these elementary inconsistencies.

Although this kind of elementary logical advantage is not something we appreciate every time we use a bar chart, it is still a real advantage. To wit, suppose someone reports on the same topic, but this time, with the "bullet slide" in Figure 65 instead of the bar chart. The bullet slide describes selected trends and values in an English text, but the set of information expressed in this slide is inconsistent. According to the third and fourth sub-bullet of the second bullet, the retention rate of part-time students is 56.2% at S institutions and 55.3% at VS institutions. This implies that an institution group with stricter admission policies has a lower retention rate for part-time students than one with less strict admission policies, contradicting the content of the first bullet of the slide.

- Groups of institutions with stricter admission policies have better rates in overall graduation, retention of part-time students, and retention of full-time students.
- Full-time students retention rate is higher than part-time student retention rate in every group of institutions:
  - 96.1 vs 82.5 % in the HS institutions,
  - 85.1 vs 55.3 % in the VS institutions,
  - 79.5 vs 56.2 % in the S institutions,
  - 75.5 vs 45.1 % in the LS institutions,
  - 61.4 vs 40.7 % in the OP institutions,

FIGURE 65 "Bullet slide" whose informational content contains a basic inconsistency that a bar chart is immune to.

This contradiction is neither the result of incorrect construction of the text in the slide nor of incorrect reading of the text. The text is perfectly well-formed and correctly read. It is just that the text is not auto-consistent with respect to even this level of elementary inconsistencies. English texts are such a flexible system of expressing information, so they allow any elementary inconsistencies to sneak into their informational content. Thus, when a set of information is expressed in the form of text, we do not get the same assurance of consistency as we obtain when it is expressed in the form of a chart. This seems to partly explain the danger of easy reliance on bullet slides in data presentation, which Edward Tufte has warned against in his booklet (Tufte 2003).[18]

In fact, the kind of basic auto-consistency illustrated here with a system of bar charts holds in many other systems of statistical charts, where the magnitudes of target objects are indicated by the magnitudes of graphical objects (e.g., the lengths of bars, the areas of pies, the distances of points from an axis). Given such a statistical chart, people are assured of the elementary consistency of whatever set of information it expresses. People do not have to be concerned about the possibility that the information set presented to them might confuse or mislead them because of some elementary inconsistency buried in it. The assurance of the elementary consistency of expressed data may not be as spectacular an advantage as the demonstrations of consistency or non-consequence that we discussed in previous subsections, but it can be

[18]Note that no bar charts can express the set of information expressed by this bullet slide. If the bar representing the part-time student retention rate of S institutions reached the 56.2% level while the bar representing the part-time student retention rate of VS institutions reached only the 55.3% level, then one of the bar groups in the chart would lack the downward shortening pattern, so the chart does not express what the first bullet of the slide says.

quite important when further cognitive tasks are founded on the data, such as when complex scientific reasoning or decision making are at stake.

4.2 Analysis

As these examples have shown, auto-consistency is a fairly general, perhaps universal, property of graphical representation systems. The fact that it confers a graphical system the consistency check potential is also well understood, thanks mainly to the work by Barwise and Etchemendy (1994; 1995a; 1995b). However, the questions still remains on the semantic mechanism *behind* an auto-consistency property of a representation system. Exactly what makes a graphical system auto-consistent? What prevents a member of a graphical system from expressing a certain range of inconsistencies?

4.2.1 Inconsistency Tracking

Before we embark on the analysis of the auto-consistency phenomenon, we need a small review of logic to see how we can characterize the inconsistency of a set of conditions using the notion of constraint. We have already indicated that a set Δ of conditions is inconsistent if it is not possible that all members of Δ hold of a single object or situation. Thus, we can express the inconsistency of Δ as the constraint $\Delta \vdash \emptyset$, where $\emptyset$ is the empty set. For, remember that $\Delta \vdash \Delta'$ means that it is impossible for all members of Δ to hold of an object or situation without some member of $\emptyset$ holding of it. Since $\emptyset$ has no member, it is impossible that some member of $\emptyset$ holds of an object or situation. Thus, it is impossible for all members of Δ to hold of an object or situation, i.e., that Δ is inconsistent. This is what $\Delta \vdash \emptyset$ means.

In the following analysis, we apply this logic to source types and target types of a representation system. We express the inconsistency of a set of types with the constraint of the form $\Theta \vdash \emptyset$, where Θ is either a set of source types or a set of target types as the case may be.

The Case of Bar Charts

Intuitively, auto-consistency involves a certain correspondence between inconsistent sets of conditions in the domain of representations and inconsistent sets of conditions in the domain of represented objects. We call this correspondence *inconsistency tracking*. In the case of the system of bar charts (Section 4.1.5), there are certain constraints on the way bars are spatially arranged. Due to these constraints, a certain arrangement of bars is just impossible, and this impossibility corresponds to the impossibility of a certain combination of the numerical values represented by the bars. More specifically, due to the system's syntactic stipulations and natural spatial constraints, the following three source types cannot hold together of a single bar chart:

(θ_{19}) Groups of institutions with stricter admission policies have better rates in overall graduation, retention of part-time students, and retention of full-time students,	(θ_{20}) The part-time student retention rate in the VS institutions is 55.3%,	(θ_{21}) The part-time student retention rate in the S institutions is 56.2%	$\vdash \emptyset$
$\Uparrow$	$\Uparrow$	$\Uparrow$	
(σ_1) Each of the groups of bars for full-time student retention rates, part-time student retention rates, and overall graduation rates has a downward shortening pattern,	(σ_2) The right end of the bar for the part-time student retention rate of VS institutions is at the 55.3% level,	(σ_3) The right end of the bar for the part-time student retention rate of S institutions is at the 56.2% level	$\vdash \emptyset$

FIGURE 66 Tracking of an inconsistency that contributes to the auto-consistency of the present system of bar charts.

(σ_1) Each of the groups of bars for full-time student retention rates, part-time student retention rates, and overall graduation rates has a downward shortening pattern.

(σ_2) The right end of a bar for the part-time student retention rate of VS institutions is at the 55.3% level.

(σ_3) The right end of a bar for the part-time student retention rate of S institutions is at the 56.2% level.

Due to the semantic conventions associated with the system of bar charts, source types σ_1– σ_3 respectively indicate the following target types:

(θ_{19}) Groups of institutions with stricter admission policies have better rates in overall graduation, retention of part-time students, and retention of full-time students.

(θ_{20}) The part-time student retention rate in the VS institutions is 55.3%.

(θ_{21}) The part-time student retention rate in the S institutions is 56.2%.

Corresponding to the inconsistency of the set $\{\sigma_1, \sigma_2, \sigma_3\}$, these indicated target types cannot hold together of a single situation. That is, the set $\{\theta_{19}, \theta_{20}, \theta_{21}\}$ is also inconsistent. Figure 66 depicts this situation schematically.

Now, σ_1, σ_2, and σ_3 are respectively the only source types in this system

$$\begin{array}{ccc} \Delta & \vdash & \emptyset \\ \Uparrow & & \\ \Gamma & \vdash & \emptyset \end{array}$$

FIGURE 67 Tracking of an inconsistency in a representation system.

that indicate θ_{19}, θ_{20}, and θ_{21}. Thus, in order for a bar chart in the system to express θ_{19}– θ_{21}, the conditions σ_1– σ_3 must hold of it. Yet σ_1– σ_3 cannot hold together of any single bar chart. It follows that no bar charts in the system can express the conditions θ_{19}– θ_{21}.

In Figure 66, we see an inconsistent set of conditions ($\{\theta_{19},\theta_{20},\theta_{21}\}$) tracked by another inconsistent set of conditions ($\{\sigma_1,\sigma_2,\sigma_3\}$) through the semantic relation $\Rightarrow$ of the system. This correspondence accounts for the inability of the system of bar charts to express a particular inconsistent set of information. The system obviously involves many other cases of inconsistency tracking, and they combine to define a significant range of inconsistent sets of information that cannot be expressed in the system.

The General Case

We can generally characterize an inconsistency tracking of a representation system in the following way.

Definition 5 (inconsistency tracking) Let Δ be a set of target types of a representation system $\mathcal{R}$.

1. System $\mathcal{R}$ *induces* Δ *as inconsistent* if there is a set Γ of source types of $\mathcal{R}$ such that:
   (a) Γ is the unique set of source types of $\mathcal{R}$ such that $\Gamma \Rightarrow \Delta$,
   (b) $\Gamma \vdash \emptyset$.
2. System $\mathcal{R}$ *tracks an inconsistency* Δ if $\mathcal{R}$ induces Δ as inconsistent and Δ is in fact inconsistent (i.e., $\Delta \vdash \emptyset$).

Figure 67 shows this relationship schematically. As we have illustrated with the example of the system of bar charts, a representation system $\mathcal{R}$ cannot express any inconsistency Δ it tracks. Generally, for a representation d of $\mathcal{R}$ to express Δ, there must be a set Γ of source types such that (i) all its members hold together of d and (ii) $\Gamma \Rightarrow \Delta$. However, the system has at most one such set Γ of source types (by clause 1a above), and this set Γ is inconsistent itself (by clause 1b). This implies that its members cannot hold together of d.

Typically, if a system ever tracks an inconsistency, it tracks more than one. Every representation system cited in Section 4.1 tracks more than one inconsistency, as should be clear from the discussions of each example. So, it makes sense to talk about the *range* of inconsistencies tracked by a given system $\mathcal{R}$. The notion of auto-consistency is defined relative to this range of inconsistent sets of information:

Definition 6 (Auto-Consistency) Given a representation system $\mathcal{R}$, we use $\bot_{\mathcal{R}}$ to refer to the range of inconsistent sets of information that $\mathcal{R}$ tracks. We say $\mathcal{R}$ is *auto-consistent over* $\bot_{\mathcal{R}}$.

Since $\mathcal{R}$ cannot express any inconsistency that it tracks, it can express no member of $\bot_{\mathcal{R}}$. Thus, if a set of information is expressed in R, it is a proof that the set of information is outside the range $\bot_{\mathcal{R}}$ of inconsistencies, namely, a proof of its *consistency relative to* $\bot_{\mathcal{R}}$. A proof thus obtained is often taken for granted, but it can bring about a significant advantage in solving inferential problems. It allows one to confirm the realizability of one's design ideas (e.g., cross-section diagrams and furniture maps in Subsection 4.1.1), to filter out information that does not follow from the given information (e.g., Hyperproof diagrams in Subsection 4.1.2 and geometry diagrams in Subsection 4.1.3), and to read off a set of information with assurance of consistency (e.g., bar charts in Subsection 4.1.4).

Generally, the more inconsistencies $\mathcal{R}$ tracks, the more significant is this proof of consistency. For if $\mathcal{R}$ tracks more inconsistencies, it can provide a proof of consistency relative to a larger $\bot_{\mathcal{R}}$, indicating that a set of information is immune to the larger set $\bot_{\mathcal{R}}$ of inconsistencies. If, on the other hand, $\mathcal{R}$ tracks only a small $\bot_{\mathcal{R}}$ of inconsistencies, it can prove only a basic consistency of a set of information, the immunity to a small range of inconsistencies.

4.2.2 Application to Specific Cases

Let us see how the general analysis works to account for specific cases of auto-consistency.

The Case of Cross-Section Diagrams

Remember that the system of cross-section diagrams discussed in Subsection 4.1.1 is auto-consistent over geometrically impossible arrangements of physical objects. We considered the following set of conditions as an example of a geometrically impossible arrangement:

(θ_1) Terrace B is horizontal and 24-foot deep.

(θ_2) Terrace C is horizontal and 24-foot deep.

(θ_4) The north ends of terraces B, C, and D make a $\frac{5}{24}$ slope with its south end down.

(θ_5) The north end of B is connected to the south end of A with a vertical wall of 3 feet.

The system of cross-section diagrams cannot express this set of informa-

(θ_1) Terrace *B* is horizontal and 24-foot deep,	(θ_2) Terrace *C* is horizontal and 24-foot deep,	(θ_4) The north ends of terraces *B*, *C*, and *D* make a $\frac{5}{24}$ slope with its south end down,	(θ_5) The north end of *B* is connected to the south end of *A* with a vertical wall of 3 feet	$\vdash \emptyset$
$\Uparrow$	$\Uparrow$	$\Uparrow$	$\Uparrow$	
(σ_4) A line representing *B* is horizontal and 12-inches wide,	(σ_5) A line representing *C* is horizontal and 12-inches wide,	(σ_6) The right ends of lines for *B*, *C*, and *D* make a $\frac{5}{24}$ slope with its left end down,	(σ_7) The right end of a line for *B* is connected to the left end of a line for *A* with a line of 1.5 inches	$\vdash \emptyset$

FIGURE 68 Tracking of an inconsistency that contributes to the auto-consistency of the system of cross-section diagrams.

tion because the unique set of source types that collectively indicates this set is inconsistent itself. That is, the following set of source types is inconsistent:[19]

(σ_4) A line representing *B* is horizontal and 12-inch wide.

(σ_5) A line representing *C* is horizontal and 12-inch wide.

(σ_6) The right ends of lines for *B*, *C*, and *D* make a $\frac{5}{24}$ slope with its left end down.

(σ_7) The right end of a line for *B* is connected to the left end of a line for *A* with a vertical line of 1.5 inches.

Thus, the system tracks the inconsistency $\{\theta_1, \theta_2, \theta_4, \theta_5\}$, and that accounts for its inability to express this set of information. Figure 68 schematically shows the inconsistency tracking in the present case.

As elements of cross-section diagrams (e.g., points, lines, angles) are geometrical objects themselves, the system tracks a wide range of geometrically impossible arrangements of objects in this way. Thus, we can use the system as a probe for geometrical consistency, just as Quist and Petra did in the example in Subsection 4.1.1.

[19] As with the description of target types, all indications of length and slopes in the following descriptions of source types are approximate. Also, it is customary in a sketch drawing like this to use several lines to approximate a single line, but we treat such a set of lines as though they were a single line, identifying those lines with the line they approximate.

(θ_9)	(θ_{10})	(θ_{11})	(θ_{12})	(θ_{14})	
Blocks *b* and *c* are in the same row,	Blocks *a* and *d* are in the same row,	Blocks *d* and *f* are in the same row,	Block *a* is to the left of block *b*,	Blocks *a* and *f* are not in the same row	$\vdash \emptyset$
$\Uparrow$	$\Uparrow$	$\Uparrow$	$\Uparrow$	$\Uparrow$	
(σ_8)	(σ_9)	(σ_{10})	(σ_{11})	(σ_{12})	
Icons labeled "*b*" and "*c*" are in the same row,	Icons labeled "*a*" and "*d*" are in the same row,	Icons labeled "*d*" and "*f*" are in the same row,	An icon labeled "*a*" is to the left of an icon labeled "*b*,"	Icons labeled "*a*" and "*f*" are not in the same row	$\vdash \emptyset$

FIGURE 69 Tracking of an inconsistency that contributes to the auto-consistency of the system of Hyperproof diagrams.

The Case of Hyperproof Diagrams

In Subsection 4.1.2, we saw that the system of Hyperproof diagrams cannot express the set consisting of the following pieces of information:

(θ_9) Blocks *b* and *c* are in the same row.

(θ_{10}) Blocks *a* and *d* are in the same row.

(θ_{11}) Blocks *d* and *f* are in the same row.

(θ_{12}) Block *a* is to the left of block *b*.

(θ_{14}) Blocks *a* and *f* are not in the same row.

The system cannot express this set of information because the unique set of source types collectively indicating this set is itself inconsistent. Figure 69 shows this relationship of inconsistency tracking.

In this way, the system of Hyperproof diagrams tracks many inconsistent sets of conditions that violate the constraints in the blocks world. It fact, it tracks *every* such inconsistent set, as far as it consists of target types that the system can express individually. In the terminology introduced in Subsection 4.2.1, this is a case where the set $\perp_{\mathcal{R}}$ of inconsistencies over which $\mathcal{R}$ is auto-consistent is maximal relative to its expressive coverage. Thus, the consistency proved by the existence of a Hyperproof diagram is strong, with no possibility that it may be inconsistent relative to some larger range of inconsistencies. Note that such a complete tracking of inconsistencies is achieved partly because the blocks world, being an artificial world for logic exercises, has a rather simple set of constraints and partly because the system of Hyperproof diagrams is tailor-made to express information about this world.

The situation is different when the target of a representation system is real-world situations, where constraints of multiple origins determine what is possible and what is impossible. Remember that the system of cross-section diagrams discussed in Subsection 4.1.1 tracks only those inconsistencies rooted in geometrical constraints on two-dimensional planes. It leaves out other inconsistencies rooted in such factors as three-dimensionality, soil conditions, and environmental protection.

The situation is different also when the syntactic and semantic conventions of a system are not "tailor-made" to a particular target domain. For example, the general syntactic-semantic conventions of bar-charts are adopted in various occasions to represent many different kinds of situations. As such, they are not exactly fitted to any specific one of these domains, tracking only some of the inconsistencies holding in it. As we have seen with our sample bar-chart system in Subsection 4.1.5, they do track elementary inconsistencies such as value-inconsistency, violation of asymmetry, and violation of transitivity, but special inconsistencies rooted in the social or economical circumstances of the relevant educational institutions must be left out of tracking. The same holds for the system of Euler diagrams discussed in Subsection 4.1.4, as it is the specific application of the general syntactic and semantic conventions of Euler diagrams to a particular domain of people and their occupations. Set-theoretic inconsistencies are tracked by those general conventions, but domain-specific constraints rooted in the nature of the represented group of people and occupations are necessarily left out.

The Case of Geometry Diagrams and Euler Diagrams

Similar accounts apply to the auto-consistency of geometry diagrams (Subsection 4.1.3) and of Euler diagrams (Subsection 4.1.4). We invite the reader to examine each system, come up with a set of inconsistent target types that the system of geometry diagrams cannot express, and apply the definition of inconsistency tracking to explain why it cannot do so.

4.3 Questions and Responses

Let us consider, as usual, some of the natural questions that would arise as to our analysis and the key concepts in it. Four questions are picked out.

4.3.1 Question: Are They Really Auto-Consistent?

Contrary to what you say in Subsection 4.1.4, a Euler diagram can express a set-theoretically inconsistent set of information. Figure 70 is one of such Euler diagrams. To be sure, you assume that a system of Euler diagram usually comes with the syntactic stipulation that the same name should not label more than one circle, and under this constraint, Figure 70 would not count as a possible diagram. This constraint however appears to be an artificial, arbitrary stipulation that you make just to support your argument. The system of

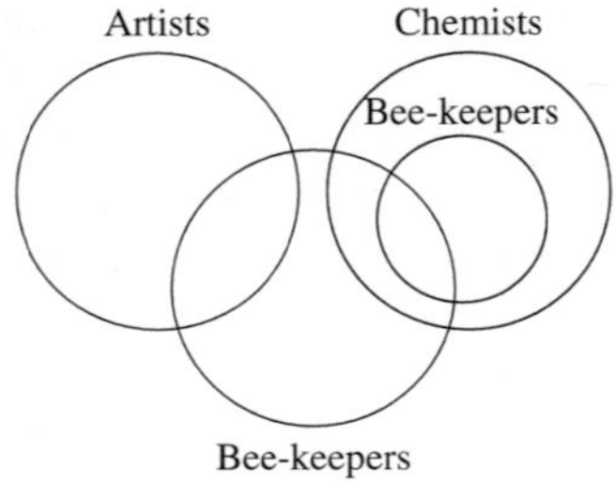

FIGURE 70 Alleged Example of an Euler diagram that expresses an inconsistent set of information.

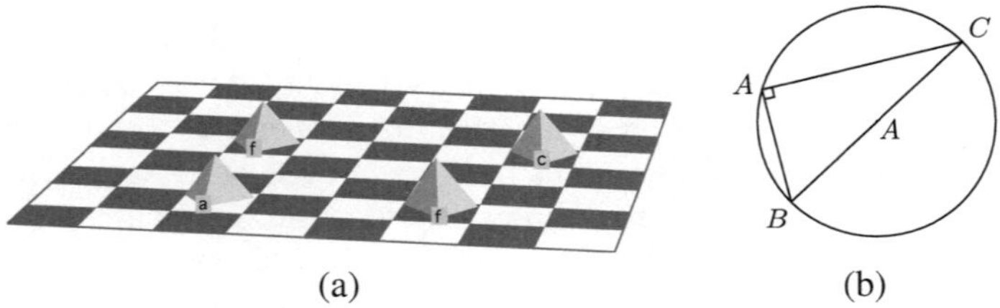

FIGURE 71 Alleged example of (a) a Hyperproof diagram and (b) a geometry diagram that express inconsistent sets of information.

plain Euler diagrams, without syntactic stipulations of this kind, is much less constrained and is not auto-consistent over set-theoretic inconsistencies.

Moreover, once we abandoned the assumption of artificial syntactic stipulations, other systems of diagrams cited in Section 4.1 are not auto-consistent over the ranges of inconsistencies that you claim they are. For example, Figure 71-a is a Hyperproof diagram that expresses a set of conditions that violates the constraints on the blocks world, and Figure 71-b is a geometry diagram that expresses a geometrically inconsistent set of conditions.

Response

No syntactic stipulations mentioned in this book are creations of the author. As we discussed earlier in our response to a different question (Section 3.3.1), a representation system is a set of regularities according to which a group of people are engaged in a particular representational practice. A syntactic stipulation is one of such regularities, so what syntactic stipulations are included in a given representation system is an objective matter that a theorist need investigate, rather than just assume.

In this light, our example of Euler diagrams points to an important function of syntactic stipulations in general. Because of a syntactic stipulation, a Euler diagram with the same name labeling multiple circles cannot be drawn, and under this constraint, no diagram in the system can express a set-theoretic inconsistency. This limitation confers the system the potential of consistency check , and that explains why people have used and kept using this syntactic stipulation.

The same explanation applies to the syntactic stipulations in the systems of Hyperproof diagrams and of geometry diagrams. In fact, there is a common pattern in the syntactic stipulations used by all three systems considered here: they all stipulate that the same name does not label more than one object. A syntactic stipulation of this kind makes the system a "token-referential system" (Stenning and Inder 1995: p. 320), where an object in a diagram refers to a unique object in the represented situation. Stenning and Inder (1995) pointed out that token-referentiality is a hallmark of a diagrammatic representation system as distinguished from linguistic representation systems. In this connection, our discussion has shown the way a syntactic stipulation contributes to token-referentiality of a representation system, which in turn contributes to auto-consistency of the system that implies clear cognitive advantages.

Syntactic stipulations are essentially arbitrary and should be distinguished from natural constraints (e.g., transitivity or asymmetry of certain spatial relations) that *necessarily* regulate spatial configurations of diagrams. Yet, just because they are arbitrary, it does not follow that they are without function. The above discussion shows that they are instrumental to an important semantic property of a representation system, and as such, they are an integral part of a representation system. Evaluated with these syntactic stipulations taken into account, the systems discussed in Section 4.1 are genuinely auto-consistent.

4.3.2 Question: Difference from the Standard Counter-Example Method?

You emphasized the importance of non-consequence proofs that consists in the drawing of a counter-example diagram (Sections 4.1.2–4.1.4). Is the method fundamentally different from the standard, language-based counter-example method, where a counter-example is described by a set of sentences? After all, diagrams are just syntactic objects just as sentences are, and the two methods are not different in using these syntactic objects to represent a possible situation in which the negation of the purported consequence does not hold while the premises hold.

Response

Any counter-example method, whether language-based or diagram-based, needs to show (1) that the situation that it presents as a counter-example is a possible situation, and (2) that the negation of the purported conclusion,

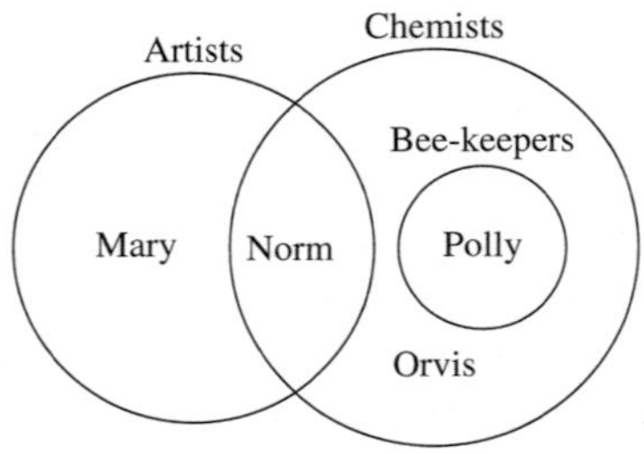

FIGURE 72 Euler diagram that demonstrates the non-consequence of information θ_{18} from premises θ_{15}–θ_{17}. Shown again.

along with all premises, holds of that situation. The fundamental difference between the two methods resides in the way (1) is shown.

Going back to the case of Euler diagrams in Subsection 4.1.4, consider how one would show the non-consequence of θ_{18} from θ_{15}-θ_{17} in the language-based method.

(θ_{15}) There are at least one artist, chemist, and bee-keeper.

(θ_{16}) No artists are bee-keepers.

(θ_{17}) All bee-keepers are chemists.

(θ_{18}) No artists is a chemist.

In particular, how would one go about describing in English the counter-example situation depicted in the Euler diagram in Figure 72? The following is one of the descriptions that one may come up with:

> May, Norm, Orvis, Polly, and nobody else are in the room. May is an artist but neither a chemist nor a bee-keeper. Norm is both an artist and chemist but not a bee-keeper. Orvis is a chemist but neither an artist nor a bee-keeper. Polly is both a bee-keeper and a chemist but not an artist.

Imagine we are given only this text (without the Euler diagram), and we try to figure out whether this situation is possible. What guarantees that this is a possible situation? Nothing but the content of the text itself. So, you need to figure out the content of this text before you can make a judgement as to the possibility of the described situation.

Contrast this to the case where we are given the Euler diagram in Figure 72 as the depiction of a counter-example. Once we see that it is a well-formed Euler diagram, we do not have to ponder if it describes a possible situation. *The system of Euler diagrams guarantees that it is a possible situation.* To be more specific, the system of Euler diagrams has a set of general syntactic stipulations that, combined with natural spatial constraints, regulate the way circles and names are spatially configured in an Euler diagram, and it guarantees that any diagram constructed in accordance with this set of syntactic stipulations expresses a consistent set of information. Syntax plus natural constraints guarantees the consistency of semantic content, and importantly, it is a *system-wide* guarantee.

In the case of the English text, there is no such system-wide guarantee for the consistency of its semantic content. The presenter of the counter-example must adjust the semantic contents of individual sentences to maintain consistency. Accordingly, the receiver must ascertain consistency by looking into the contents of individual sentences, not just by confirming the text's conformity to a set of general syntactic stipulations. The process of showing and checking (1) is thus fundamentally different between the language-based and the diagram-based method of counter-example.

The process of showing and checking (2) is not fundamentally different, however. To see if the premises plus the negation of the purported consequence hold of the represented situation, we need look into the semantic content of the given representation, whether it is a text or a diagram. The only notable difference is that, due to the nature of its strategy, the diagram-based method presents a diagram that expresses all relevant information (the premises and the negation of the purported conclusion) in itself, while the text presented in the language-based counter-example method tends to express only a part of the relevant information and let the reader infer the rest. Although potentially important, this difference will not be further investigated here.

4.3.3 Question: Unsoundness in Inconsistency Tracking?

In Subsection 4.2.1 and other places, you emphasized that the inconsistency tracking done by a representation system can be limited, in the sense that it may track just some of the inconsistencies holding in the target domain, with other inconsistencies left untracked. Using a term introduced in Subsection 4.2.1, this is the case where the system fails to induce some sets of target types as inconsistent, although they are in fact inconsistent. Figure 73-a depicts this situation of incomplete tracking. Now, the question is if the opposite can hold, where a system induces a set of target types as inconsistent, even though it is in fact consistent. Figure 73-b shows this case of unsound tracking. If a representation system is not sound in inconsistency tracking, it is incapable of expressing a genuinely possible set of target types, even though the set consists of target types that are individually expressible in the system.

Response

Unsound tracking of an inconsistency is certainly a logical possibility, but representation systems that actually do it are difficult to find. The situation is similar to the one discussed back in Subsection 2.4.4, where we find that

$$\begin{array}{ccc} \Delta \vdash \emptyset & \qquad & \Delta \nvdash \emptyset \\ \Uparrow & & \Uparrow \\ \Gamma \nvdash \emptyset & & \Gamma \vdash \emptyset \\ \text{(a)} & & \text{(b)} \end{array}$$

FIGURE 73 Comparison of (a) incomplete tracking of an inconsistency to (b) unsound tracking of an inconsistency.

almost no real-life diagrammatic representation systems do unsound tracking of a consequential constraint, although it is logically possible. These two phenomena seem to stem from the same root, but we do not have any systematic account to offer, except the conjecture that it has to do with a unified way in which many primitive diagrammatic systems are conceived.

4.3.4 Question: Diagrammatic Proof of Inconsistency?

If you are right in the observation that a typical diagrammatic system is sound in inconsistency tracking, it means that *every* set of target types that a system induces as inconsistent is in fact inconsistent. In that case, is it not possible to prove the *inconsistency* of a set of information on the basis of the fact that the system induces it as inconsistent? Take the system of Euler diagrams cited above. The unique set of source types that collectively indicates the set $\{\theta_{15}, \theta_{16}, \theta_{17}, \theta_{38}\}$ of target types is $\{\sigma_{13}, \sigma_{14}, \sigma_{15}, \sigma_{16}\}$:

(θ_{15}) There are at least one artist, chemist, and bee-keeper.

(θ_{16}) No artists are bee-keepers.

(θ_{17}) All bee-keepers are chemists.

(θ_{38}) All chemists are artists.

(σ_{13}) There are circles labeled "Artists," "Chemists," and "Bee-keepers."

(σ_{14}) A circle labeled "Artists" is outside of a circle labeled "Bee-keepers."

(σ_{15}) A circle labeled "Bee-keepers" is enclosed in a circle labeled "Chemists."

(σ_{16}) A circle labeled "Chemists" is enclosed in a circle labeled "Artists."

Now it is impossible that source types σ_{13}, σ_{14}, σ_{15}, and σ_{16} hold together of a single Euler diagram. Thus, this system of Euler diagram induces the set $\{\theta_{15}, \theta_{16}, \theta_{17}, \theta_{38}\}$ as inconsistent. Given that this system is sound in inconsistency tracking, this fact should suffice as proof of the inconsistency of $\{\theta_{15}, \theta_{16}, \theta_{17}, \theta_{38}\}$.

Taking another example, the system of Hyperproof diagrams induces the set $\{\theta_9, \theta_{10}, \theta_{41}, \theta_{42}\}$ of information as inconsistent.

(θ_9) Blocks b and c are in the same row.

(θ_{10}) Blocks a and d are in the same row.

(θ_{41}) Block a is in front of block b.

(θ_{42}) Block d is in back of block c.

Specifically, the set $\{\sigma_{17}, \sigma_{18}, \sigma_{19}, \sigma_{20}\}$ of source types is inconsistent, while it is the unique set of source types that collectively indicates $\{\theta_9, \theta_{10}, \theta_{41}, \theta_{42}\}$.

(σ_{17}) Icons labeled "*b*" and "*c*" are in the same row.

(σ_{18}) Icons labeled a and d are in the same row.

(σ_{19}) An icon labeled "*a*" is in front of an icon labeled "*b*."

(σ_{20}) An icon labeled "*d*" is in back of an icon labeled "*c*."

Assuming the system is sound in inconsistency tracking, this fact should be sufficient to establish the inconsistency of $\{\theta_9, \theta_{10}, \theta_{41}, \theta_{42}\}$.

Generally, this "diagrammatic method" of proving inconsistency should be applicable to any set Δ of information as long as $\mathcal{R}$ is sound in inconsistency tracking and Δ is in the range $\perp_{\mathcal{R}}$ of inconsistencies tracked by $\mathcal{R}$.

Response

Agreed. You can prove the inconsistency of a set of information in this way. Unlike the case of proving consistency, however, such proof would involve more than just drawing a single diagram, for in this case, the key fact to establish is the *inconsistency* of a set of source types, as opposed to its consistency.

To be more specific, let us see how we would demonstrate that σ_{13}, σ_{14}, σ_{15}, and σ_{16} cannot hold together of a single Euler diagram. Since one cannot show this fact by drawing a Euler diagram in which these source types all hold, the most plausible strategy is to draw a Euler diagram in which only *some* members of this set hold, and see that the diagram cannot be modified to have the other members hold while keeping the original members holding of it. Figure 74 shows a Euler diagram in which the member σ_{14} of the set $\{\sigma_{13}, \sigma_{14}, \sigma_{15}, \sigma_{16}\}$ holds. With this diagram at hand, it is relatively easy to see that we cannot modify this diagram to have the other members σ_{13}, σ_{15}, and σ_{16} hold while keeping σ_{14} holding of it. Since a circle labeled "Chemists" can appear only once in a Euler diagram in this system, source types σ_{15} and σ_{16} require us to enclose the circle labeled "Bee-keeper" in a circle labeled "Chemists" while having the latter enclosed in the circle labeled "Artists."[20] Meeting this requirements at the same time is obviously impossible.

[20]"Enclosing" here means that one circle has no area outside of the other. So, two circles completely overlapping is a case of enclosing.

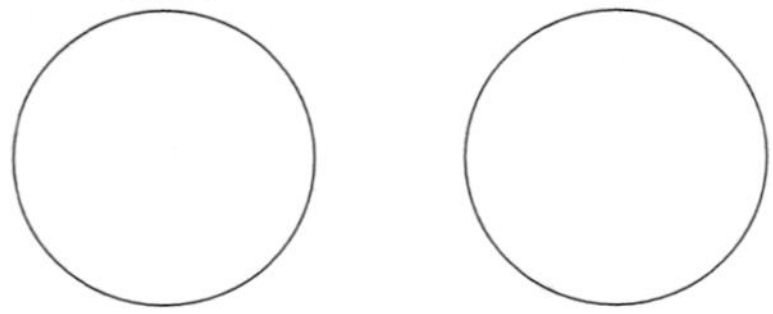

FIGURE 74 Euler diagram that helps demonstrate the inconsistency of the set $\{\sigma_{13}, \sigma_{14}, \sigma_{15}, \sigma_{16}\}$ of source types.

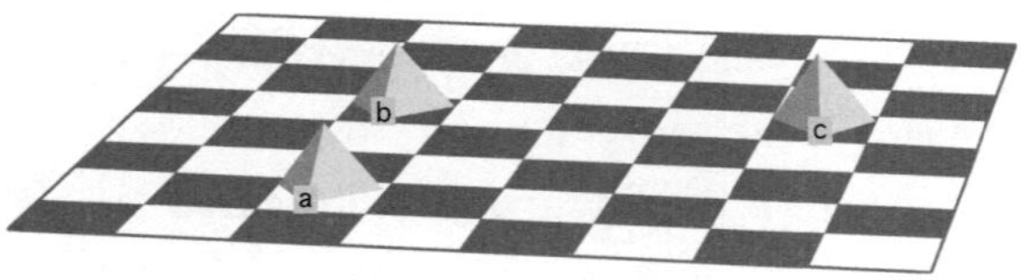

FIGURE 75 Hyperproof diagram that helps demonstrate the inconsistency of the set $\{\sigma_{17}, \sigma_{18}, \sigma_{19}, \sigma_{20}\}$ of source types.

To demonstrate that the set $\{\sigma_{17}, \sigma_{18}, \sigma_{19}, \sigma_{20}\}$ of source types cannot hold of any Hyperproof diagram, we might first draw a diagram (Figure 71) in which only σ_{17} and σ_{19} hold. Figure 75 shows such a Hyperproof diagram. Then it is obvious that we cannot place an icon labeled "*d*" in the same row as the icon labeled "*a*" while placing it in back of the icon labeled "*c*." Noting that this result does not depend on the exact initial placement, sizes, and shapes of the icons labeled "*b*," "*c*," and "*a*," you can see there is no way to arrange icons to express the set in question.

There might be other ways of demonstrating non-drawability, but what is common in this endeavor is the fact that we use diagrams as "inferential surrogates" for the situations they represent. Remember that our original task in each of the above examples is to demonstrate the inconsistency of a set of target types. Because our strategy is to demonstrate this by demonstrating that a representation system induces the set as inconsistent, our main task was changed to demonstrate the inconsistency of a set of source types. Thus, instead of trying to see the inconsistency of a set of conditions on the sets of chemists, bee-keepers, and artists, we try to establish the inconsistency of a certain spatial arrangement of *circles* and *names* on a Euler diagram. In the case of Hyperproof diagrams, we try to establish the inconsistency of a certain formation of block *icons* on the *picture* of a chess board picture instead of trying to see the inconsistency of a formation of blocks on a chess board.

As the diagrams used as surrogates in this process are perceptually accessible, they accept such operations as envisaging and spatial analysis transfer, which we will discuss later in Chapter 6. Although the exact mechanism of these "surrogate inferences" is yet to be investigated from both logical and empirical points of view, inconsistency inferences based on such visual surrogates can have quite distinctive cognitive characteristics compared with purely mental inconsistency inferences that deal with the problem situation directly.

4.4 Chapter Takeaways

- An auto-consistency is a property of a representation system, where a certain range of inconsistent sets of information cannot be expressed.
- Many diagrammatic representation systems are auto-consistent over a significant range of inconsistencies. Among such systems are the systems of cross-section diagrams, Hyperproof diagrams, geometry diagrams, Euler diagrams, bar charts, and other statistical charts.
- An auto-consistency holds in a given diagrammatic representation system when, due to constraints on spatial configurations of representing objects, the unique set of source types that collectively indicates an inconsistent set of information is itself inconsistent, and when this tracking of inconsistency is accumulated to cover a significant range of inconsistent set of information.
- When a representation system is auto-consistent over a range of inconsistent sets of information, the consistency of a set of information relative to this range can be confirmed by constructing a representation in the system that expresses all members of that set of information.
- The limited consistency thus established also can be used to confirm, in the correspondingly limited sense, that a piece of information is not a consequence of a set of premises.

5

Potential for Derivative Meaning

Graphical systems can support meaning relations that go beyond their basic semantic conventions. This enables us to read off a richer set of information directly from graphics, expanding the expressive coverage of a system with a relatively simple set of basic semantic conventions.

5.1 The Phenomenon

5.1.1 Tufte (1983): Scatter Plots

At the outset of his well-known book, *The Visual Display of Quantitative Information*, Tufte (1983) cited scatter plots (originally from Anscombe 1973) as examples of statistical charts with clear "graphical excellence" (p. 13). Figure 76 has the four scatter plots he cited, where plots I–IV correspond to the data sets I–IV in Table 1.

On the one hand, each of these scatter plots is "inter-translatable" with the table of its corresponding letter—Plot I, for example, provides sufficient information to reproduce data set I, and vice versa. In this limited sense, each plot has the same informational content as its corresponding table.

On the other hand, there is a definite sense in which each plot expresses more information than the corresponding table does. The particular shape formed by the dots on a scatter plot seems to indicate some general fact about the data, or more precisely, about the situation that the data are about. Thus, the particular shape in Figure 76-I indicates that the variables X and Y are in a certain degree of linear correlation, the shape in Figure 76-II that they have a high degree of curvilinear correlation, the shape in Figure 76-III that X and Y are near-proportional except for a single outlier, and the shape in Figure 76-IV that X is constant except for a single outlier.

In fact, it is precisely because of this additional informational relation that we sometimes prefer scatter plots to simple tables in displaying quantitative data. Kosslyn (1994) clarifies this function of scatter plots:

> Scatter plots...employ point symbols (such as dots, small triangles, or squares)

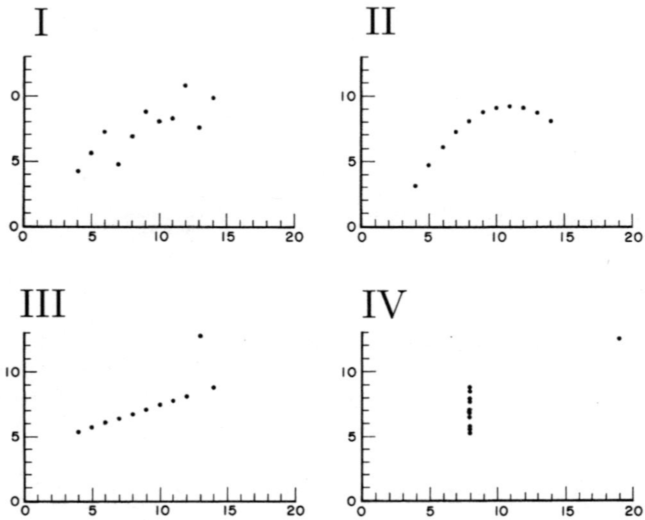

FIGURE 76 Scatter plots expressing data sets A–D in Table 1. Reprinted from Anscombe (1973) with permission of the American Statistical Association (http://www.amstat.org). The regression lines shown in the original plots are removed in republication.

> as content elements. The height of each point symbol indicates an amount. These displays typically include so many points that they form a cloud; information is conveyed by the shape and the density of the cloud. (Kosslyn 1994: p. 46.)

It is clear, and almost trivial, that the particular shape and density of the cloud formed in some scatter plots indicate information that the corresponding table of data would not. In fact, this kind of additional informational relations are quite prevalent in graphical representation systems, and as the above example illustrates, their existence is often the very reason why a given system is more effective than others as a method of displaying certain sets of information.

Nevertheless, when it comes to the question of how such additional informational relations arise in given representation systems, things are much less clear.

The basic semantic conventions for the present system of scatter plots is most likely the following:

(1) If a dot appears at Y-coordinate *m* and X-coordinate *n*, then there is an

TABLE 1 Four sets of data represented by scatter plots in Figure 76.

I		II		III		IV	
X	Y	X	Y	X	Y	X	Y
10	8.04	10	9.14	10	7.46	8	6.58
8	6.95	8	8.14	8	6.77	8	5.76
13	7.58	13	8.74	13	12.74	8	7.71
9	8.81	9	8.77	9	7.11	8	8.84
11	8.83	11	9.26	11	7.81	8	8.47
14	9.96	14	8.10	14	8.84	8	7.04
6	7.24	6	6.13	6	6.08	8	5.25
4	4.26	4	3.10	4	5.39	19	12.50
12	10.84	12	9.13	12	8.15	8	5.56
7	4.82	7	7.26	7	6.42	8	7.91
5	5.68	5	4.74	53	5.73	8	6.89

instance in the data with the X- value and the Y-value represented by *n* and *m*, respectively.

Clearly, (1) is not a natural constraint in the environment. It holds because a group of people started to conform to it at some time in history and kept respecting it for the common interest of effective communication. It is a *convention* in this sense. Apparently, there are no other conventions from which (1) can be derived. It is a *basic* convention in this sense.

The informational relation that we saw between the shapes of dots and the general facts indicated by them have a different logical and historical status. Take, for example, the following instance of an informational relation that underlies the interpretation of plots discussed earlier.

(2) If the dots make a cluster along a line with a left-up slant, it indicates that variables *X* and *Y* are linearly correlated to that extent.

It is not necessary for the establishment of this instance of an informational relation for the relevant group of people to start to conform to a new basic semantic convention. Constraint (2) seems to be something that naturally holds once the basic convention (1) is established. In this sense, it is a *derived* constraint. The same applies to the other cases, listed more precisely as (3)–(5).

(3) If the dots make a cluster along a quadratic curve, it indicates that variables *X* and *Y* are in a quadratic correlation to the extent of clustering.

(4) If the dots make a cluster along a line with a left-up slant and there is a single dot far apart from the cluster, it indicates that variables *X* and *Y* are correlated to the extent of clustering and that the tendency has an outlier to the extent of apartness.

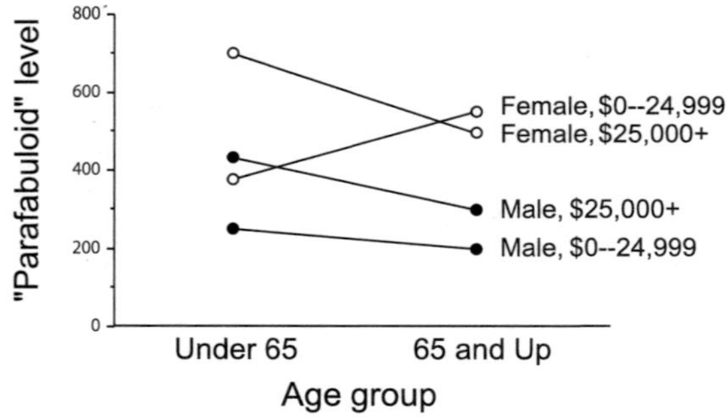

FIGURE 77 Line graph analogous to the one discussed by Kosslyn (1994).

(5) If the dots make a cluster along a vertical line with a single dot far apart from the cluster, it indicates that X is a constant variable to the extent of clustering and that the tendency has an outlier to the extent of apartness.

These are not part of basic semantic conventions of the system. Rather, they are part of an informational relation derived from the basic semantic conventions.

5.1.2 Kosslyn (1994): Line Graphs

Just as dots in a scatter plot form a particular cloud that carries information, data points in a line graph form an informative slope or curve when connected by lines. For example, Kosslyn (1994) explained the utility in the following way, referring to a line graph similar to Figure 77 :

> Here it is apparent that all groups but one show lower levels of "parafabuloid" with increased age—women in the lower income group have the reverse trend. Spotting this trend in the table is difficult, but seeing in this graph is easy: Differences in the orientations of the lines convey the different trends, and the eye and mind quickly register such differences. (Kosslyn 1994: p. 10.)

Kosslyn is appealing to the following instances of informational relations supported by this particular system of line graphs:

(6) If all lines but one line are slanted right-down, it means that all groups of people but one have lower parafabuloid levels at ages 65 and up than at ages under 65.

It is due to (6) that the line graph lets us "spot" the relevant trend of parafabuloid levels.

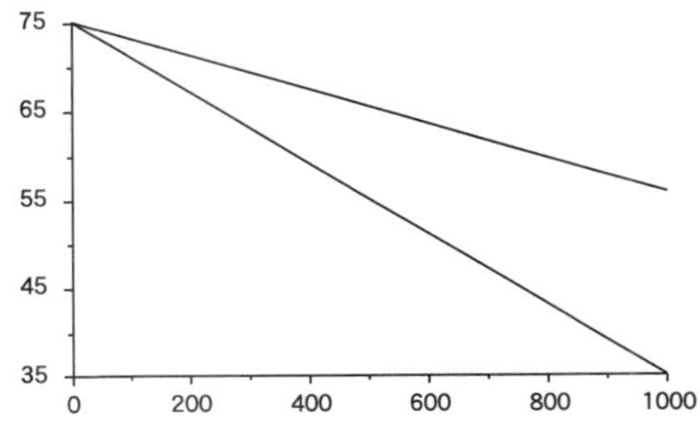

FIGURE 78 Line graph depicting functional dependencies of temperature and altitude in atmosphere under two different conditions. Reprinted with permission from Gattis and Holyoak (1996) (American Psychological Association).

Yet (6) is an instance of an informational relation that we would call "derivative." The basic semantic conventions for line graphs are what can be summarized in the following way, where *n* is the parafabuloid level indicated by the *Y*-coordinate *N* on the plot:

(7) If a line tagged with sex *S* and income *I* has its left end at *Y*-coordinate *N*, it means that the parafabuloid level of people under age 65 with sex *S* and income *I* is *n*.

(8) If a line tagged with sex *S* and income *I* has its right end at *Y*-coordinate *N*, it means that the parafabuloid level of people of age 65 or up with sex *S* and income *I* is *n*.

Clearly, (6) is not part of the basic semantic conventions summarized in (7) and (8). Rather, it is a relation that naturally holds once the basic semantic conventions are established.

Yet the derivative informational relation illustrated above significantly contributes to the overall informational utility of line graphs. In Kosslyn's words, "Patterns of lines can signal specific information for readers who have had experience with similar line graphs (and so have appropriate knowledge)" (Kosslyn 1994: p. 34).

Gattis and Holyoak (1996) used different kinds of line graphs for their experiments on people's interpretations of slopes made by lines. Figure 78 shows one of their stimulus graphs.

Here, two lines, one solid and the other dotted, represent the functional dependencies of temperature and altitude in the atmosphere under two different conditions. The vertical positions of each line at selected horizontal positions specifically indicate the temperature values at the corresponding altitudes in the given case. The basic semantic conventions are thus concerned with individual vertical positions at which the relevant line crosses individual horizontal positions.

Of course, these point-wise facts are not the only features in the line graph that carry useful information. For example, the solid line in Figure 78 makes a steeper right-down incline than the dotted line. This fact, which compares the inclines made by two lines, means that the temperature decreases "faster" under the condition represented by the solid line than that represented by the dotted line. This is not part of the basic semantic conventions that we have just specified.

For that matter, the sheer fact that each line is straight means that there is a linear dependency in each case. If one of the lines were rectilinear, it would mean an abrupt change in the temperature at a certain altitude. Further, if two lines inclined in opposite directions, say one right-up and the other right down, it would mean that there were reverse trends in the two cases. All these appear to fall into the category of derivative meaning.

5.1.3 Pinker (1990): Bar Charts

Bar charts are another interesting example. Consider Figure 79, reproduced from Pinker (1990), which represents the price of "graphium" over the first six months of a year. Here the horizontal position of a bar stands for a particular month of the year, and the bar's height stands for a particular price. Thus, the appearance of a bar of a particular height at a particular horizontal position indicates the price of graphium in a particular month. This much is the basic semantic conventions of this system of bar charts. However, this bar graph lets us see (much) more than the prices of graphium in individual months. For example, the block of bars at the positions from "March" to "June" has the shape of a "descending staircase" (Pinker 1990), and this means that the price of graphium steadily declined from March to June. Clearly, the informational relation illustrated here is *not* part of the basic semantic conventions we have just specified. The basic semantic conventions are only concerned with the meanings carried by the heights of individual bars, not the information carried by the shapes of blocks of bars.

Also, observe that the bar for March is *taller* than that for January. This means that the price of graphium is higher in March than in January, but strictly speaking, this informational relation is not part of the basic semantic conventions. It is concerned with a *pair of bars* and comparisons of their lengths, and not the absolute heights of individual bars. In the same vein, the fact that the bar for March is the tallest in all bars, or the fact that the bar for April is the second tallest, bear meanings that are beyond the immediate scope of the basic semantic conventions.

Some bar charts even use their overall shapes to indicate more abstract facts about the represented data. Histograms, which are a special kind of bar charts, are a well-known example. At the basic level, they use the heights of individual bars to indicate the frequencies with which specific values or

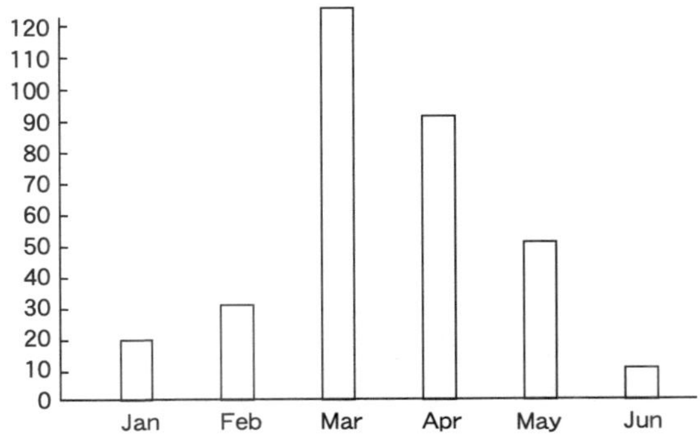

FIGURE 79 Bar chart plotting prices of "graphium" over the first six months of a year. Reprinted with permission of Taylor and Francis Group LLC Books, from Pinker (1990); permission conveyed through Copyright Clearance Center, Inc.

value ranges are instantiated in a set of data. At a more abstract level, the extent to which these bars form a symmetrical bell shape indicates the extent to which the data are normally distributed. The extent to which the bars form a curve skewed to the left (right) indicates the extent to which the data are biased toward the lower (higher) range, and the extent to which the bars form a spike-like curve with long tails to both sides indicates the extent to which the values are concentrated around the mean value. All these are familiar to students of statistics, but the informational relation illustrated here is not part of the basic semantic conventions for histograms.

Since Playfair's invention of line graphs, numerous kinds of statistical data graphs have been invented and inherited (See Tufte (1983) and Tufte (1990) for excellent compilations). Star graphs, pie graphs, box charts, stem-and-leaf charts, and combinations of these are all kinds of statistical graphs. Further, many, perhaps most, of them support some kinds of derivative meanings that we have illustrated thus far. This has led some practitioners and researchers to distinguish the "levels of meaning" carried by statistical graphs (Bertin, 1967/1973,1977/1981; Wainer, 1992; Guthrie et al., 1993), where derivative meanings in our sense are separated from basic, conventional meanings. Practitioners commonly call higher-level meanings the "main messages" of statistical graphs, while Bertin (1977/1981) goes so far as to characterize their communication as the "purpose" of preparing statistical graphs.

5.1.4 Kinnear and Wood (1997): Contour Maps

The phenomenon of derivative meaning is not confined to the display of numerical data in statistical charts. It is also relevant when we display spatial information in the form of maps. Kinnear and Wood (1987) used a *topographic contour map*, such as the one in Figure 80, for their experiments on the relationship between geographical expertise and recognition memory with maps. Here, every point on the map plane is projected onto a particular position on the mapped terrain, and every contour curve is associated with a certain altitude. The basic semantic conventions for contour lines then state that if a contour line passes a point on the map plane, it means that the position to which the point is projected has the altitude associated with the contour line.

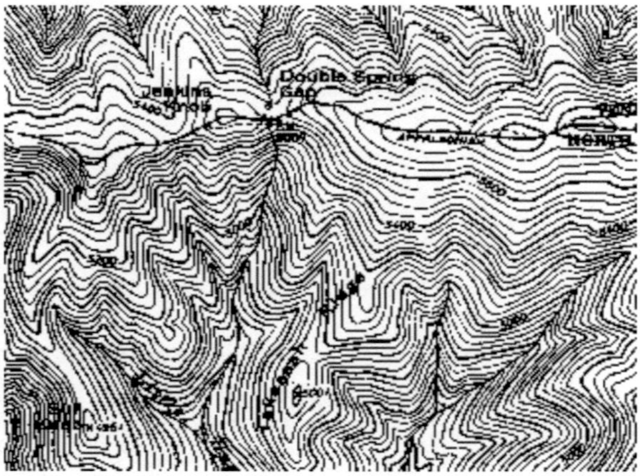

FIGURE 80 Topographic contour map. North Carolina, Map Collection, 1935–, courtesy of the U. S. Geological Survey.

While the basic semantic conventions are thus concerned with individual points on individual contour lines, patterns formed by several proximate contour lines also carry important information about the mapped terrain. When contour lines are close together in a mapped area, it means that the corresponding geographical area makes a steep slope, whereas when they are distant, it means that the area is a shallow slope; when contour lines for ascending altitudes form a group of near-cocentric closed curves, it means there is a mountain peak in the area represented by the inner-most closed curve. There are also general patterns of contour curves that mean the shapes of spurs, valleys, cirques, and other landforms. In fact, the existence of such meaning relations is a presupposition for some psychological studies of reading skills and memory for contour maps (Kinnear and Wood 1987; Gilhooly et al. 1988).

A similar derivative meaning relation is naturally attributed to meteorological maps, where isobars replace contour lines in contour maps. Lowe (1989, 1994) emphasized the importance of appreciating perceptual patterns of isobars (“secondary structures” in his terminology) for the skill of reading meteorological maps. One of the example he cites is concerned with the pattern

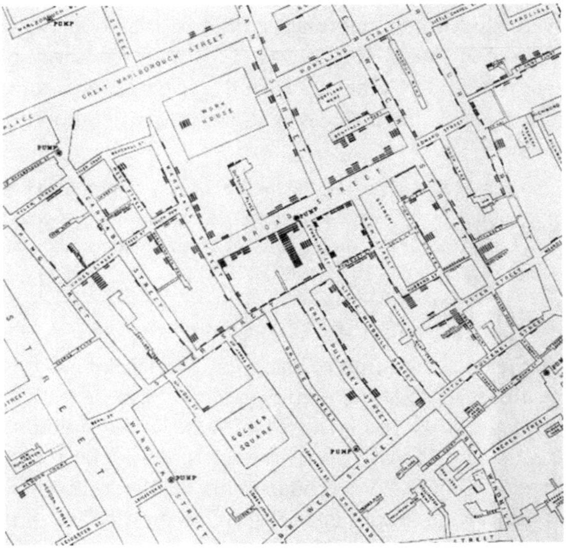

FIGURE 81 A part of Snow's data map. Reproduced from Snow (1855).

that indicates "a trough structure", where several adjacent isobars indicates "localised dips or bumps" that constitute a trough axis.

5.1.5 Tufte (1983): Data Maps

The shapes formed by small symbols carry important information when they are overlaid on a map. Tufte (1983) cites Dr. John Snow, who used a street map of central London to plot the locations of deaths from cholera in September 1854 (Snow 1855). The individual short bars on his map in Figure 81 indicate the locations of deaths, while the neighborhood pumps are indicated by small double circles labeled "PUMP."

As with the case of dots on a scatter plot, the bars on the map form a pattern of a specific shape and density, and it clearly indicates a general pattern of a spatial distribution of cholera deaths. Specifically, the fact that many bars appear around the pump symbol on the band denoting the Broad Street carries information that many deaths occurred around the water pump on Broad Street.

Nevertheless, the informational relation between these two facts is derivative. The basic semantic conventions consist of the usual semantic rules for the map, along with the specifications for the meanings of the dots and crosses at particular locations on the map. They do not include the informational relation from the concentration of dots to that of deaths.

This derivative informational relation has historically played a crucial role in the actual use of this map. The creator of the map, Dr. Snow, used the informational relation to find that the pump on Broad Street was contaminated. He had the handle of the pump removed and ended the neighborhood epidemic. Tufte (1983) says:

> Of course the link between the pump and the disease might have been revealed by computation and analysis without graphics, with some good luck and hard work. But, here at least, graphical analysis testifies about the data far more efficiently than calculation. (Tufte 1983: p. 24.)

5.1.6 Olivier (2001): Node-Edge Graphs

Olivier (2001) discussed a simple, but clear example of what we would call derivative meanings in node-edge graphs. Generally, the basic semantic rule for a node-edge graph is that if an edge connects two nodes, it means that the objects denoted by the two nodes are in the particular relationship represented by the edge. Figure 82, reproduced from Olivier (2001), is an example of a directed node-edge graph, and an arrow from the "C" node to the "E" node means, for example, that object C is related to object E under the binary relation represented by the edge.

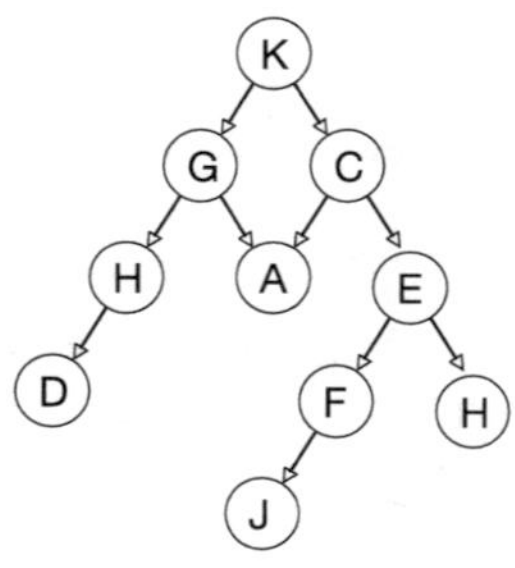

FIGURE 82 A node-edge graph. Reprinted from Olivier (2001) with permission of Springer; permission conveyed through Copyright Clearance Center, Inc.

In Figure 82, we see a line of connected nodes from the node labeled "K" down to the node labeled "D". Now the simple fact that node labeled "H" belongs to this line of nodes means that object K is an ancestor of object H in the binary relation represented by this node-edge graph. One may directly extract this ancestor relation from the membership of the "H" node in the node group, without knowing how other members of the group are connected by individual edges. The informational relation illustrated here is clearly beyond the basic semantic rule for node-edge graphs, which only allows one to observe a pair-wise relation of objects from the presence or absence of an edge between a pair of nodes.

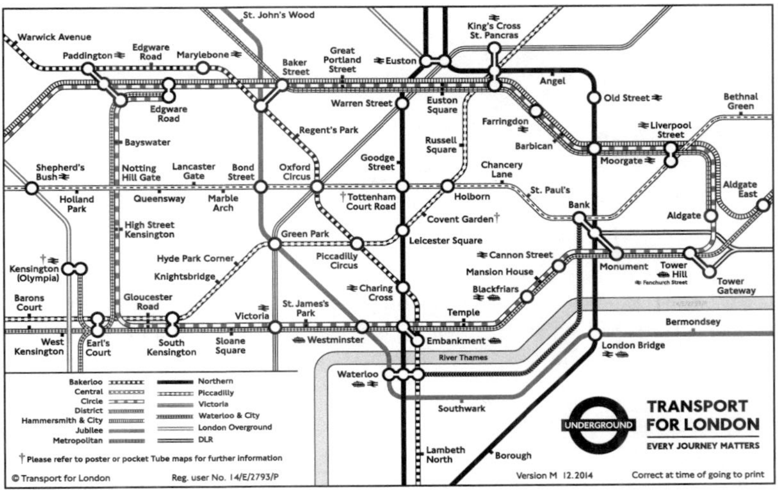

FIGURE 83 A version of London's underground map. Reproduced on a license agreement.

For that matter, the presence of a *path* between two nodes carries a meaning beyond the scope of the basic semantic conventions. Consider, for example, a route map of the underground system in London, or of the subway system of any reasonably large city. Although such a route map typically gives hints about the geographical locations of stations on the map, it is essentially a (non-directed) node-edge graph. Now, one version of London's underground map (Figure 83) contains a "triplet" path (|||), consisting of several consecutive edges, from the "Victoria" node to the "Warren Street" node, and it means that Victoria station is connected to Warren Street station by the Victoria line. Note that this informational relation goes beyond the basic semantic conventions specified above: one refers to the meaning of a *path* consisting of multiple edges connecting two station nodes, while the other refers to the meaning of a single edge connecting two station nodes.

The London Underground map seems to support numerous other instances of derivative informational relation. One often counts the number of station nodes along a particular path between two station nodes, and the result means the number of stations in the corresponding portion of the represented tube line. Moreover, if there are a *greater number of* station nodes along a path than another between two station nodes, it means there are a *greater number of* stations along one tube route than another between the corresponding stations. Furthermore, the number of nodes directly connected to a particular node indicates the number of stations directly connected to the station represented

	USA	CHN	KOR	BRA	TUR	SRB
USA	-	○	○	○	○	○
CHN	●	-	○	●	○	○
KOR	●	●	-	○	●	○
BRA	●	○	●	-	○	○
TUR	●	●	○	●	-	○
SRB	●	●	●	●	●	-

(a)

	USA	CHN	KOR	BRA	TUR	SRB	W	L	PTS
United States	-	○ 3-0	○ 3-1	○ 3-1	○ 3-0	○ 3-0	5	0	15
China	● 0-3	-	○ 3-2	● 2-3	○ 3-1	○ 3-1	3	2	9
South Korea	● 1-3	● 2-3	-	○ 3-0	● 2-3	○ 3-1	2	3	8
Brazil	● 1-3	○ 3-2	● 0-3	-	○ 3-2	○ 3-0	3	2	7
Turkey	● 0-3	● 1-3	○ 3-2	● 2-3	-	○ 3-0	2	3	6
Serbia	● 0-3	● 1-3	● 1-3	● 0-3	● 0-3	-	0	5	0

(b)

FIGURE 84 Round-robin tables reporting the results of a preliminary round of women's volleyball at the 2012 London Olympic games: (a) simplified version for discussion and (b) more realistic version likely to be seen in news report.

by the node, giving a scale with which we can detect the "hub" of the city's tube system.

One may also think of examples of derivative informational relations in other kinds of node-edge graphs. Specifically, Gurr and Toulas' (2000) example of the "safety" property expressed in sequential function charts, Pinker's (1990) example of the "dominance" relation expressed in linguists' phrase structure trees, and Petre and Green's (1992) example of "overview effects" of electronic circuit diagrams may fall into this category (although detailed examinations of each example would be required to verify these claims).

5.1.7 Further Examples: Round-Robin Tables

The round-robin table in Figure 84-a presents the results of a preliminary round of women's volleyball at the 2012 London Olympic games. A round-robin table typically seen in a news report has more numerical information attached to it, and looks more like Figure 84-b. We nevertheless use Figure 84-a as our example, both to simplify our discussion and highlight the characteristics of purely graphical aspects of round-robin tables.

Since this system of round-robin tables is designed to report the results of games played by a particular set of six teams (Group B) in the preliminary round, it has the following syntactic stipulations: (i) any table has six rows each headed by a name of a participating team, (ii) no two rows are headed by the same name, and no two names head a single row, (iii) there are also six columns with unique, non-repeating names, and (iv) a hyphen "-" appears at the intersection of a row and a column headed by the same name.

The system's basic semantic conventions are as follows:

(9) If a white circle appears at the intersection of a row headed by name *X* and a column headed by name *Y*, it means that the team with name *X* has won against the team with name *Y*.

(10) If a black circle appears at the intersection of a row headed by name *X* and a column headed by name *Y*, it means that the team with name *X* has lost to the team with name *Y*.

In the table in Figure 84-a, for example, a white circle appears at the intersection of the row headed by name "CHN" and the column headed by name "TUR," so we find that the China team won against the Turkey team[21]; also, a black circle appears at the intersection of the row headed by "CHN" and the column headed by "BRA," so we find that the China team lost to the Brazil team. This way of reading the table is directly legitimized by the basic semantic conventions of the system.

There are, once again, other legitimate ways of reading this table. We can see three white circles appearing in the row headed by "BRA," and we interpret them to mean that the Brazil team won against three teams. We can also obtain the same information by reading the table vertically, noticing three black circles in the column headed by "BRA." We may notice that the row headed by "USA" has a greater number of white circles than any other row, and we interpret it to mean that that the US team won against a greater number of teams than any other team.

Note that the meaning relation underlying these reading practices differs from that directly legitimized by the basic semantic conventions given in (9) and (10). For comparison, (11)–(13) explicitly display the relevant part of the meaning relation in question:

(11) If three white circles appear in a row headed by "BRA," it means that the Brazil team has won against three teams.

(12) If three black circles appear in a column headed by "BRA," it means that the Brazil team has won against three teams.

(13) If more white circles appear in a row headed "CHN" than in a row headed "TUR," it means that the China team has won against more teams than the Turkey team has.

(14) If a row headed by "USA" has a greater number of white circles than any other row, it means that the US team has won against a greater number of teams than any other team.

Here, (11)–(14) are concerned with an entire row, an entire column, or multiple rows, specifying the information carried by the particular distributions of white and black circles on them. In contrast, the basic semantic conventions (9) and (10) are concerned with individual cells, specifying the information carried by individual symbols appearing on them. If the basic se-

[21] We follow a convention adopted in many news reports and use three-letter country names (e.g., "KOR," "BRA," "SRB") to denote the national teams representing the relevant countries.

mantic conventions are cell-wise meaning specifications, (11)–(14) are row-wise, column-wise, or multi-row-wise meaning specifications. The system of round-robin tables supports a meaning relation that goes beyond what is directly legitimized by its basic semantic conventions.

5.2 Analysis

We have seen several examples of graphical systems that support what may be called "derivative meaning," which is the additional informational relation derivable from, but not included in, the system's basic semantic conventions. We will now look into the details of the logical relation between derivative meaning and basic semantic conventions.

5.2.1 Abstraction Tracking

We will start with the analysis of a case of derivative meaning in the system of round-robin tables. This example will help us identify the general property of *abstraction tracking* as the logical origin of derivative meaning.

The Case of Round Robin Tables

Remember that the system of round-robin tables has the following basic semantic conventions:

(15) If a white circle appears at the intersection of a row headed by name X and a column headed by name Y, it means that the team with name X has won against the team with name Y.

(16) If a black circle appears at the intersection of a row headed by name X and a column headed by name Y, it means that the team with name X has lost to the team with name Y.

To make the description of our analysis easier, let us use the following abbreviations to denote the source types and target types mentioned in the above conventions:

$\circ(X, Y)$	=	a white circle appears at the intersection of a row headed by name X and a column headed by name Y (source type)
$\bullet(X, Y)$	=	a black circle appears at the intersection of a row headed by name X and a column headed by name Y (source type)
$\mathrm{W}(x, y)$	=	team x has won against team y (target type)
$\mathrm{L}(x, y)$	=	team x has lost to team y (target type)

Thus, for example, ●("BRA", "KOR") is the source type where a black circle appears at the intersection of a row headed by name "BRA" and a column headed by name "KOR," and L(BRA, KOR) is the target type where the Brazil team has lost to the South Korea team.

Using these abbreviations, and assuming x and y are the teams denoted by names X and Y, we can rewrite the basic semantic conventions (15) and (16) in the following way:

(17) $\circ(X, Y)$ means $\mathrm{W}(x, y)$

(18) $\bullet(X, Y)$ means $\mathrm{L}(x, y)$

As we have explained back in Section 2.2.4, a semantic convention in a representation system is a *regularity* in the local environment of the users of the system. In the present case, this means that the following two general constraints hold as semantic conventions (17) and (18):

(19) If $\circ(X, Y)$ holds, then $\mathrm{W}(x, y)$ holds.

(20) If $\bullet(X, Y)$ holds, then $\mathrm{L}(x, y)$ holds.

This conception of meaning as constraints plays an important role in our account.[22]

We have seen several cases where a legitimate meaning relation holds between pairs of source and target types that are not directly specified by the system's basic semantic conventions. The following is one such example:

(21) If three white circles appear in a row headed by "BRA," it means that the Brazil team has won against three teams.

Let us take up this instance and investigate where its legitimacy comes from. We use the following abbreviations to denote the source type and target type involved in this instance of derivative meaning relation:

$\circ_n$(X-row)	=	n white circles appear in a row headed by name X (source type)
W_n(x)	=	team x has won against n teams (target type)

With these abbreviations, our question is how a valid informational relation holds from $\circ_3$("BRA"-row) to W_3(BRA).

Note that $\circ_3$("BRA"-row) is an *abstract* condition, in the sense that there are several alternative ways in which it is true. One way in which $\circ_3$("BRA"-row) is true is that white circles appear at the intersections of a row headed by "BRA" with columns headed by "CHN," "TUR," and "SRB," while black circles appear in its intersections with columns headed by "USA" and "KOR."

[22]We adopt this conception from situation semantics (Barwise and Perry 1983/1999), in which it is used to analyze how linguistic meaning is embedded in the general information flows in the environment. This conception is also an application of Lewis's general theory of convention (Lewis 1975/1985).

Using the abbreviations introduced above, we model this "way" as the following set Γ_1 of source types:

$$\Gamma_1 = \left\{ \begin{array}{ll} \circ(\text{"BRA","CHN"}), & \circ(\text{"BRA","TUR"}), \\ \circ(\text{"BRA","SRB"}), & \bullet(\text{"BRA","USA"}), \\ \bullet(\text{"BRA","KOR"}) & \end{array} \right\}$$

What exactly does it mean to say Γ_1 is a "way" in which $\circ_3$("BRA"-row) holds? In our conception, it means that if all members of Γ_1 hold of any round-robin table in the present system, then $\circ_3$("BRA"-row) necessarily holds of that table. Thus, it is an assertion of a *constraint* holding on possible properties of round-robin tables. The constraint is based on the system's syntactic formation rules, plus natural spatial constraints on table structures. Generally, to state that a set Γ of source types is a way in which $\circ_3$("BRA"-row) holds is to state that the following constraint holds:

(22) If all members of Γ hold, then $\circ_3$("BRA"-row) holds.

The following is another particular way in which $\circ_3$("BRA"-row) holds:

$$\Gamma_2 = \left\{ \begin{array}{ll} \circ(\text{"BRA","CHN"}), & \circ(\text{"BRA","TUR"}), \\ \bullet(\text{"BRA","SRB"}), & \circ(\text{"BRA","USA"}), \\ \bullet(\text{"BRA","KOR"}) & \end{array} \right\}$$

Note that set Γ_2 is a way in which $\circ_3$("BRA"-row) holds, despite the fact that Γ_2 is not satisfied by the particular table in Figure 84-a. This is so because $\circ_3$("BRA"-row) necessarily holds of a round-robin table *provided* all members of Γ_2 hold of it.

Keeping on with this enumeration $\Gamma_1, \Gamma_2, \ldots$, we eventually obtain the collection of sets of source types that exhausts all the alternative ways in which $\circ_3$("BRA"-row) holds. This collection, which we call $\mathcal{G}$, contains ${}_5C_3 = 10$ sets. Figure 85 lists the entire collection $\mathcal{G}$.

Since each member of $\mathcal{G}$ is a particular way in which $\circ_3$("BRA"-row) holds, the following relation holds between $\circ_3$("BRA"-row) and $\mathcal{G}$:

(23) For every set Γ in $\mathcal{G}$, if all members of Γ hold, then $\circ_3$("BRA"-row) holds.

Conversely, since $\mathcal{G}$ exhausts all the ways in which $\circ_3$("BRA"-row) is true, the following relation also holds:

(24) If $\circ_3$("BRA"-row) holds, then there is some set Γ in $\mathcal{G}$ whose members all hold.

To succinctly express the bi-directional implications (23) and (24) between $\circ_3$("BRA"-row) and $\mathcal{G}$, we say $\circ_3$("BRA"-row) is an *abstraction over* $\mathcal{G}$, writing:

(25) $\circ_3$("BRA"-row) $\bowtie \mathcal{G}$.

$$
\mathcal{G} = \left\{
\begin{array}{ll}
\left\{ \begin{array}{ll} \circ(\text{“BRA”,“CHN”}), & \circ(\text{“BRA”,“TUR”}), \\ \circ(\text{“BRA”,“SRB”}), & \bullet(\text{“BRA”,“USA”}), \\ \bullet(\text{“BRA”,“KOR”}) & \end{array} \right\}, &
\left\{ \begin{array}{ll} \circ(\text{“BRA”,“CHN”}), & \circ(\text{“BRA”,“TUR”}), \\ \bullet(\text{“BRA”,“SRB”}), & \circ(\text{“BRA”,“USA”}), \\ \bullet(\text{“BRA”,“KOR”}) & \end{array} \right\}, \\
\left\{ \begin{array}{ll} \circ(\text{“BRA”,“CHN”}), & \circ(\text{“BRA”,“TUR”}), \\ \bullet(\text{“BRA”,“SRB”}), & \bullet(\text{“BRA”,“USA”}), \\ \circ(\text{“BRA”,“KOR”}) & \end{array} \right\}, &
\left\{ \begin{array}{ll} \circ(\text{“BRA”,“CHN”}), & \bullet(\text{“BRA”,“TUR”}), \\ \circ(\text{“BRA”,“SRB”}), & \circ(\text{“BRA”,“USA”}), \\ \bullet(\text{“BRA”,“KOR”}) & \end{array} \right\}, \\
\left\{ \begin{array}{ll} \circ(\text{“BRA”,“CHN”}), & \bullet(\text{“BRA”,“TUR”}), \\ \circ(\text{“BRA”,“SRB”}), & \bullet(\text{“BRA”,“USA”}), \\ \circ(\text{“BRA”,“KOR”}) & \end{array} \right\}, &
\left\{ \begin{array}{ll} \circ(\text{“BRA”,“CHN”}), & \bullet(\text{“BRA”,“TUR”}), \\ \bullet(\text{“BRA”,“SRB”}), & \circ(\text{“BRA”,“USA”}), \\ \circ(\text{“BRA”,“KOR”}) & \end{array} \right\}, \\
\left\{ \begin{array}{ll} \bullet(\text{“BRA”,“CHN”}), & \circ(\text{“BRA”,“TUR”}), \\ \circ(\text{“BRA”,“SRB”}), & \circ(\text{“BRA”,“USA”}), \\ \bullet(\text{“BRA”,“KOR”}) & \end{array} \right\}, &
\left\{ \begin{array}{ll} \bullet(\text{“BRA”,“CHN”}), & \circ(\text{“BRA”,“TUR”}), \\ \circ(\text{“BRA”,“SRB”}), & \bullet(\text{“BRA”,“USA”}), \\ \circ(\text{“BRA”,“KOR”}) & \end{array} \right\}, \\
\left\{ \begin{array}{ll} \bullet(\text{“BRA”,“CHN”}), & \circ(\text{“BRA”,“TUR”}), \\ \bullet(\text{“BRA”,“SRB”}), & \circ(\text{“BRA”,“USA”}), \\ \circ(\text{“BRA”,“KOR”}) & \end{array} \right\}, &
\left\{ \begin{array}{ll} \bullet(\text{“BRA”,“CHN”}), & \bullet(\text{“BRA”,“TUR”}), \\ \circ(\text{“BRA”,“SRB”}), & \circ(\text{“BRA”,“USA”}), \\ \circ(\text{“BRA”,“KOR”}) & \end{array} \right\}
\end{array}
\right\}
$$

FIGURE 85 Collection $\mathcal{G}$ of sets of conditions, which exhausts all particular ways in which $\bigcirc_3$(“BRA”-row) holds.

Intuitively, $\bigcirc_3$(“BRA”-row) captures a piece of information commonly implied by the members of $\mathcal{G}$, abstracting away the specific information particular to individual members. Hence the term “abstraction.”

Note that (25) is an abstraction relation that holds in the source domain since implications (23) and (24) are concerned with the properties of round-robin tables themselves, rather than information about the represented volleyball competition. We will now see that an abstraction relation parallel to (25) also holds in the target domain.

Just as $\bigcirc_3$(“BRA”-row) is an abstract condition, W_3(BRA) is an abstract condition. Thus, there are several alternative ways in which W_3(BRA) holds. One way is that the Brazil team wins against the China, the Turkey, and the Serbia teams but loses to the US and the South Korea teams. Using our abbreviations introduced earlier, we model this “way” as the following set of target types:

$$
\Delta_1 = \left\{ \begin{array}{ll} \text{W(BRA,CHN)}, & \text{W(BRA,TUR)}, \\ \text{W(BRA,SRB)}, & \text{L(BRA,USA)}, \\ \text{L(BRA,KOR)} & \end{array} \right\}
$$

Thus, if all members of Δ_1 hold, then W_3(BRA) necessarily holds. This is due to the game regulation specifying the set of participating teams and their matches. Here is another particular way in which W_3(BRA) holds:

$$
\Delta_2 = \left\{ \begin{array}{ll} \text{W(BRA,CHN)}, & \text{W(BRA,TUR)}, \\ \text{L(BRA,SRB)}, & \text{W(BRA,USA)}, \\ \text{L(BRA,KOR)} & \end{array} \right\}
$$

Keeping on with this enumeration, we again obtain the collection of sets of target types that exhausts all the alternative ways in which target type

$$
\mathcal{D} = \left\{ \begin{array}{ll}
\left\{ \begin{array}{ll} \text{W(BRA,CHN)}, & \text{W(BRA,TUR)}, \\ \text{W(BRA,SRB)}, & \text{L(BRA,USA)}, \\ \text{L(BRA,KOR)} & \end{array} \right\}, &
\left\{ \begin{array}{ll} \text{W(BRA,CHN)}, & \text{W(BRA,TUR)}, \\ \text{L(BRA,SRB)}, & \text{W(BRA,USA)}, \\ \text{L(BRA,KOR)} & \end{array} \right\}, \\
\left\{ \begin{array}{ll} \text{W(BRA,CHN)}, & \text{W(BRA,TUR)}, \\ \text{L(BRA,SRB)}, & \text{L(BRA,USA)}, \\ \text{W(BRA,KOR)} & \end{array} \right\}, &
\left\{ \begin{array}{ll} \text{W(BRA,CHN)}, & \text{L(BRA,TUR)}, \\ \text{W(BRA,SRB)}, & \text{W(BRA,USA)}, \\ \text{L(BRA,KOR)} & \end{array} \right\}, \\
\left\{ \begin{array}{ll} \text{W(BRA,CHN)}, & \text{L(BRA,TUR)}, \\ \text{W(BRA,SRB)}, & \text{L(BRA,USA)}, \\ \text{W(BRA,KOR)} & \end{array} \right\}, &
\left\{ \begin{array}{ll} \text{W(BRA,CHN)}, & \text{L(BRA,TUR)}, \\ \text{L(BRA,SRB)}, & \text{W(BRA,USA)}, \\ \text{W(BRA,KOR)} & \end{array} \right\}, \\
\left\{ \begin{array}{ll} \text{L(BRA,CHN)}, & \text{W(BRA,TUR)}, \\ \text{W(BRA,SRB)}, & \text{W(BRA,USA)}, \\ \text{L(BRA,KOR)} & \end{array} \right\}, &
\left\{ \begin{array}{ll} \text{L(BRA,CHN)}, & \text{W(BRA,TUR)}, \\ \text{W(BRA,SRB)}, & \text{L(BRA,USA)}, \\ \text{W(BRA,KOR)} & \end{array} \right\}, \\
\left\{ \begin{array}{ll} \text{L(BRA,CHN)}, & \text{W(BRA,TUR)}, \\ \text{L(BRA,SRB)}, & \text{W(BRA,USA)}, \\ \text{W(BRA,KOR)} & \end{array} \right\}, &
\left\{ \begin{array}{ll} \text{L(BRA,CHN)}, & \text{L(BRA,TUR)}, \\ \text{W(BRA,SRB)}, & \text{W(BRA,USA)}, \\ \text{W(BRA,KOR)} & \end{array} \right\}
\end{array} \right\}
$$

FIGURE 86 The collection $\mathcal{D}$ of sets of conditions, which exhausts all the particular ways in which W_3(BRA) holds.

W_3(BRA) holds. This collection, which we call $\mathcal{D}$, looks like Figure 86, consisting of ${}_5C_3 = 10$ sets of target types.

Since each member of $\mathcal{D}$ is a particular way in which W_3(BRA) holds, and $\mathcal{D}$ exhausts all the ways in which W_3(BRA) is true, the following bidirectional implications hold between W_3(BRA) and $\mathcal{D}$:

(26) For every set Δ in $\mathcal{D}$, if all members of Δ hold, then W_3(BRA) holds.

(27) If W_3(BRA) holds, then there is some set Δ in $\mathcal{D}$ whose members all hold.

That is, W_3(BRA) is an *abstraction over $\mathcal{D}$ in the target domain*, so we write:

(28) $W_3(\text{BRA}) \bowtie \mathcal{D}$.

Now, let us look back to Figures 85 and 86 to compare the two collections $\mathcal{G}$ and $\mathcal{D}$. There is an obvious "semantic" correspondence between them. To wit, consider the member of $\mathcal{G}$ that is listed in the top left corner of Figure 85. This is the set Γ_1 mentioned earlier. Compare this set with the member of $\mathcal{D}$ that is listed in the top left corner of Figure 86. This is set Δ_1 also mentioned earlier. Here, the two sets are in a one-to-one correspondence under the basic semantic conventions, (17) and (18), of the system of round-robin tables: (17) relates ○("BRA","CHN"), ○("BRA","TUR"), and ○("BRA","SRB") to W(BRA,CHN), W(BRA,TUR), and W(BRA,SRB) in turn, while (18) relates ●("BRA","USA") and ●("BRA","KOR") to L(BRA,USA) and L(BRA,KOR) in turn.

This is, of course, just a special case of a collective indication that we defined in Subsection 2.2.4 and have encountered several times in previous chapters. However, to emphasize that the correspondence between Γ_1 and Δ_1 only depends on the system's *basic* semantic conventions (17) and (18), we say that Γ_1 *collectively indicates* Δ_1 *by basic semantic conventions.* Given our assumption that semantic conventions create constraints in the local environment, this implies that if all source types in Γ_1 hold, then all target types in Δ_1 must also hold.

Now, as we can see by comparing Figures 85 and 86, the entire collections $\mathcal{G}$ and $\mathcal{D}$ are closely related under this collective indication relation. In fact:

(29) $\mathcal{G}$ and $\mathcal{D}$ are in a one-to-one correspondence under the collective indication by the system's basic semantic conventions.

Figure 87 provides an overview of the relationship between $\bigcirc_3$("BRA"-row), W_3(BRA), $\mathcal{G}$, and $\mathcal{D}$. The symbol ⇂⇂ with the tag "1-1" stands for a one-to-one correspondence under the collective indication by basic semantic conventions. Here, $\bigcirc_3$("BRA"-row) abstracts over collection $\mathcal{G}$ (bottom $\bowtie$), and this abstraction relation "tracks" the abstraction relation between W_3(BRA) and $\mathcal{D}$ (top $\bowtie$). This tracking of abstraction is mediated by the one-to-one correspondence from $\mathcal{G}$ to $\mathcal{D}$ under the collective indication by the system's basic semantic conventions.

This structure of abstraction tracking accounts for the valid informational relation from source type $\bigcirc_3$("BRA"-row) to target type W_3(BRA). To wit, suppose source type $\bigcirc_3$("BRA"-row) holds. Then, by the fact $\bigcirc_3$("BRA"-row) $\bowtie$ $\mathcal{G}$, there is some set Γ in $\mathcal{G}$ whose members all hold. Since $\mathcal{G}$ and $\mathcal{D}$ are in a one-to-one correspondence, set Γ collectively indicates some set in $\mathcal{D}$ by basic semantic conventions. Let Δ be this set. Since all members of Γ hold, all members of Δ must hold. Then, since Δ is a member of $\mathcal{D}$, and since W_3(BRA) $\bowtie$ $\mathcal{D}$, it follows that W_3(BRA) holds. Thus, we see a valid informational relation holds from source type $\bigcirc_3$("BRA"-row) to target type W_3(BRA).

Figuratively speaking, this informational relation goes from $\bigcirc_3$("BRA"-row) to $\mathcal{G}$, then from $\mathcal{G}$ to $\mathcal{D}$, and then from $\mathcal{D}$ to W_3(BRA). Therefore, by transitivity, it goes directly from $\bigcirc_3$("BRA"-row) to W_3(BRA). This informational relation is a *derived* one, based on the system's basic semantic conventions, plus the two instances of abstraction relations holding in the source domain and the target domain. Clearly, we can generalize this result and state that an informational relation holds from source type $\bigcirc_n$(X-row) to target type $W_n(x)$ for any x $(0 \leq x \leq 5)$.

We can similarly account for the logical origin of the other instances of derivative meaning supported by the system of round-robin tables, including:

(30) If three white circles appear in a row headed by "BRA," it means that the Brazil team has won against three teams.

(31) If three black circles appear in a column headed by "BRA," it means that the Brazil team has won against three teams.

$$
W_3(\text{BRA}) \bowtie \left\{
\begin{array}{l}
\left\{\begin{array}{ll}\text{W(BRA,CHN)}, & \text{W(BRA,TUR)},\\ \text{W(BRA,SRB)}, & \text{L(BRA,USA)},\\ \text{L(BRA,KOR)} & \end{array}\right\},
\left\{\begin{array}{ll}\text{W(BRA,CHN)}, & \text{W(BRA,TUR)},\\ \text{L(BRA,SRB)}, & \text{W(BRA,USA)},\\ \text{L(BRA,KOR)} & \end{array}\right\},\\
\left\{\begin{array}{ll}\text{W(BRA,CHN)}, & \text{W(BRA,TUR)},\\ \text{L(BRA,SRB)}, & \text{L(BRA,USA)},\\ \text{W(BRA,KOR)} & \end{array}\right\},
\left\{\begin{array}{ll}\text{W(BRA,CHN)}, & \text{L(BRA,TUR)},\\ \text{W(BRA,SRB)}, & \text{W(BRA,USA)},\\ \text{L(BRA,KOR)} & \end{array}\right\},\\
\left\{\begin{array}{ll}\text{W(BRA,CHN)}, & \text{L(BRA,TUR)},\\ \text{W(BRA,SRB)}, & \text{L(BRA,USA)},\\ \text{W(BRA,KOR)} & \end{array}\right\},
\left\{\begin{array}{ll}\text{W(BRA,CHN)}, & \text{L(BRA,TUR)},\\ \text{L(BRA,SRB)}, & \text{W(BRA,USA)},\\ \text{W(BRA,KOR)} & \end{array}\right\},\\
\left\{\begin{array}{ll}\text{L(BRA,CHN)}, & \text{W(BRA,TUR)},\\ \text{W(BRA,SRB)}, & \text{W(BRA,USA)},\\ \text{L(BRA,KOR)} & \end{array}\right\},
\left\{\begin{array}{ll}\text{L(BRA,CHN)}, & \text{W(BRA,TUR)},\\ \text{W(BRA,SRB)}, & \text{L(BRA,USA)},\\ \text{W(BRA,KOR)} & \end{array}\right\},\\
\left\{\begin{array}{ll}\text{L(BRA,CHN)}, & \text{W(BRA,TUR)},\\ \text{L(BRA,SRB)}, & \text{W(BRA,USA)},\\ \text{W(BRA,KOR)} & \end{array}\right\},
\left\{\begin{array}{ll}\text{L(BRA,CHN)}, & \text{L(BRA,TUR)},\\ \text{W(BRA,SRB)}, & \text{W(BRA,USA)},\\ \text{W(BRA,KOR)} & \end{array}\right\}
\end{array}
\right\}
$$

$$\rceil_{1\text{-}1}\rceil$$

$$
\circ_3(\text{“BRA”-row}) \bowtie \left\{
\begin{array}{l}
\left\{\begin{array}{ll}\circ(\text{“BRA”,“CHN”}), & \circ(\text{“BRA”,“TUR”}),\\ \circ(\text{“BRA”,“SRB”}), & \bullet(\text{“BRA”,“USA”}),\\ \bullet(\text{“BRA”,“KOR”}) & \end{array}\right\},
\left\{\begin{array}{ll}\circ(\text{“BRA”,“CHN”}), & \circ(\text{“BRA”,“TUR”}),\\ \bullet(\text{“BRA”,“SRB”}), & \circ(\text{“BRA”,“USA”}),\\ \bullet(\text{“BRA”,“KOR”}) & \end{array}\right\},\\
\left\{\begin{array}{ll}\circ(\text{“BRA”,“CHN”}), & \circ(\text{“BRA”,“TUR”}),\\ \bullet(\text{“BRA”,“SRB”}), & \bullet(\text{“BRA”,“USA”}),\\ \circ(\text{“BRA”,“KOR”}) & \end{array}\right\},
\left\{\begin{array}{ll}\circ(\text{“BRA”,“CHN”}), & \bullet(\text{“BRA”,“TUR”}),\\ \circ(\text{“BRA”,“SRB”}), & \circ(\text{“BRA”,“USA”}),\\ \bullet(\text{“BRA”,“KOR”}) & \end{array}\right\},\\
\left\{\begin{array}{ll}\circ(\text{“BRA”,“CHN”}), & \bullet(\text{“BRA”,“TUR”}),\\ \circ(\text{“BRA”,“SRB”}), & \bullet(\text{“BRA”,“USA”}),\\ \circ(\text{“BRA”,“KOR”}) & \end{array}\right\},
\left\{\begin{array}{ll}\circ(\text{“BRA”,“CHN”}), & \bullet(\text{“BRA”,“TUR”}),\\ \bullet(\text{“BRA”,“SRB”}), & \circ(\text{“BRA”,“USA”}),\\ \circ(\text{“BRA”,“KOR”}) & \end{array}\right\},\\
\left\{\begin{array}{ll}\bullet(\text{“BRA”,“CHN”}), & \circ(\text{“BRA”,“TUR”}),\\ \circ(\text{“BRA”,“SRB”}), & \circ(\text{“BRA”,“USA”}),\\ \bullet(\text{“BRA”,“KOR”}) & \end{array}\right\},
\left\{\begin{array}{ll}\bullet(\text{“BRA”,“CHN”}), & \circ(\text{“BRA”,“TUR”}),\\ \circ(\text{“BRA”,“SRB”}), & \bullet(\text{“BRA”,“USA”}),\\ \circ(\text{“BRA”,“KOR”}) & \end{array}\right\},\\
\left\{\begin{array}{ll}\bullet(\text{“BRA”,“CHN”}), & \circ(\text{“BRA”,“TUR”}),\\ \bullet(\text{“BRA”,“SRB”}), & \circ(\text{“BRA”,“USA”}),\\ \circ(\text{“BRA”,“KOR”}) & \end{array}\right\},
\left\{\begin{array}{ll}\bullet(\text{“BRA”,“CHN”}), & \bullet(\text{“BRA”,“TUR”}),\\ \circ(\text{“BRA”,“SRB”}), & \circ(\text{“BRA”,“USA”}),\\ \circ(\text{“BRA”,“KOR”}) & \end{array}\right\}
\end{array}
\right\}
$$

FIGURE 87 Tracking of abstraction mediated by basic semantic conventions of the system of round-robin tables. The symbol $\rceil\rceil$ with the tag “1-1” stands for a one-to-one correspondence under the collective indication by basic semantic conventions. This relationship accounts for the valid informational relation from source type $\circ_3$(“BRA”-row) to target type W_3(BRA).

(32) If more white circles appear in a row headed “CHN” than in a row headed “TUR,” it means that the China team has won against more teams than the Turkey team has.

(33) If a row headed by “USA” has a greater number of white circles than any other row, it means that the US team has won against a greater number of teams than any other team.

We invite the reader to pick some of these to try out our account. We included our analysis of (32) in Appendix B. The analysis involves a “short-cut” for identifying the abstraction tracking relation responsible for the meaning derivation, and is of independent interest.

The General Case

The analysis of the case of round-robin tables identifies the abstraction tracking property as the logical origin of a derivative meaning relation. To characterize this property of a graphical system more generally, we first define what it is for one type to be an *abstraction* over a collection of sets of types.

Definition 7 (Abstraction) A type α *abstracts over* a collection $\mathcal{P}$ of sets of types, written $\alpha \bowtie \mathcal{P}$, if the following bi-directional constraints hold: (1) for every set Φ in $\mathcal{P}$, if all members of Φ hold, then α holds, and (2) if α holds, then there is some set Φ in $\mathcal{P}$ whose members all hold.

Note that an abstraction relation can hold in either the source domain or the target domain of a representation system. That is, type α in the above definition can be a source type with $\mathcal{P}$ being a collection of sets of source types, or α can be a target type with $\mathcal{P}$ being a collection of sets of target types.

We earlier have discussed how a basic semantic convention of a representation system constitutes a local constraint from a source type σ to a target type θ of the system. We will say σ *indicates* θ *by basic semantic conventions*, writing $\sigma \rightharpoonup \theta$, if a constraint holds from σ to θ as part of the system's semantic conventions.

The indication relation $\rightharpoonup$ is part of the larger indication relation $\Rightarrow$ for a representation system (see below for an account of why it is only part). We know what it is for a set of source types to collectively indicate a set of target types through $\Rightarrow$. The following characterizes a special case of collective indication.

Definition 8 (Collective indication by basic semantic conventions) A set Γ of source types *collectively indicates* a set Δ of target types *by basic semantic conventions* if Γ and Δ are in a one-to-one correspondence under the indication relation $\rightharpoonup$.

With these subsidiary concepts at hand, we can now define the relation of abstraction tracking in the following way:

Definition 9 (Abstraction Tracking) A source type σ *tracks* a target type θ *in abstraction* if there are a collection $\mathcal{G}$ of sets of source types and a collection $\mathcal{D}$ of sets of target types such that (i) $\sigma \bowtie \mathcal{G}$, (ii) $\theta \bowtie \mathcal{D}$, and (iii) $\mathcal{G}$ and $\mathcal{D}$ are in a one-to-one correspondence under the collective indication by basic semantic conventions.

Figure 88-a presents the general picture of the abstraction tracking relation in a representation system $\mathcal{R}$.

The immediate, and most important consequence of this definition is that whenever σ tracks θ in abstraction, a valid informational relation holds from source type σ to target type θ. We have already seen this in connection to the specific case of round-robin tables. The general argument runs as follows. By definition, abstraction relation $\sigma \bowtie \mathcal{G}$ implies (34) and abstraction relation $\theta \bowtie \mathcal{D}$ implies (35) below:

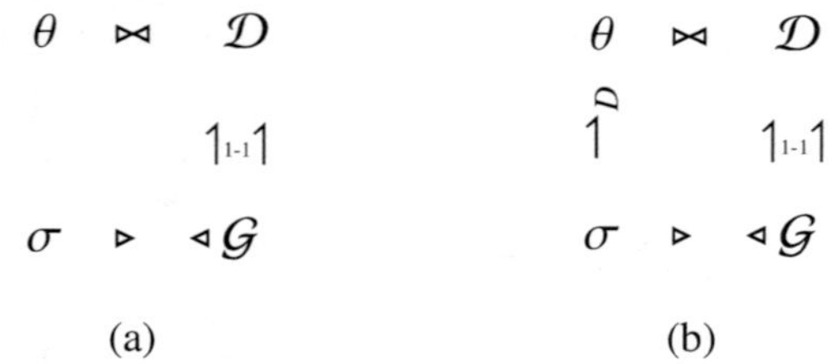

FIGURE 88 Abstraction tracking in a representation system that underlies the derivative meaning relation from source type σ to target type θ. The symbol 1 1 with the tag "1-1" stands for a one-to-one correspondence under the collective indication by basic semantic conventions. The derivative indication relation $\rightharpoonup_D$ in (b) holds whenever the abstraction tracking relation in (a) holds.

(34) If σ holds, then there is some set Γ in $\mathcal{G}$ whose members all hold.

(35) For every set Δ in $\mathcal{D}$, if all members of Δ hold, then θ holds.

Given the conception that a basic semantic convention gives rise to a local constraint, the one-to-one correspondence from $\mathcal{G}$ to $\mathcal{D}$ implies:

(36) For every set Γ in $\mathcal{G}$, if all members of Γ hold, then there is some set Δ in $\mathcal{D}$ whose members all hold.

Combining (34), (35), and (36), we find that if σ holds, then θ holds.

Therefore, we can consider a source type σ as *indicating* a target type θ whenever σ and θ stand in an abstraction tracking relation. Figure 88-b shows this situation. Unlike the indication relation $\rightharpoonup$, this indication relation includes instances that go beyond the system's basic semantic conventions. These instances make up a derivative indication relation, and we write this as $\rightharpoonup_D$. Our main proposal is that the additional meaning relation often observed in a graphical representation system is a derivative indication relation $\rightharpoonup_D$ in this sense.

On the basis of this analysis, we can clarify the nature of the indication relation $\Rightarrow$ that we have encountered in the previous chapters. It is the combination of the indication relation $\rightharpoonup$ directly established by the system's basic semantic conventions and the derivative indication relation $\rightharpoonup_D$ that we have just characterized. Thus, whenever the analysis in a previous chapter stated $\sigma \Rightarrow \theta$, it meant that either $\sigma \rightharpoonup \theta$ or $\sigma \rightharpoonup_D \theta$.[23]

[23]We can actually *identify* $\Rightarrow$ with the indication relation defined by the abstraction tracking relation. Note that the basic indication relation $\rightharpoonup$ is also part of the indication relation defined by the abstraction tracking relation. For if $\sigma \rightharpoonup \theta$, then the following trivially holds: (i) $\sigma \bowtie \{\{\sigma\}\}$, (ii) $\theta \bowtie \{\{\theta\}\}$, and (iii) $\{\{\sigma\}\}$ and $\{\{\theta\}\}$ are in a one-to-one correspondence under the collective indication by basic semantic conventions. Therefore, by the definition of $\rightharpoonup_D$, the indication

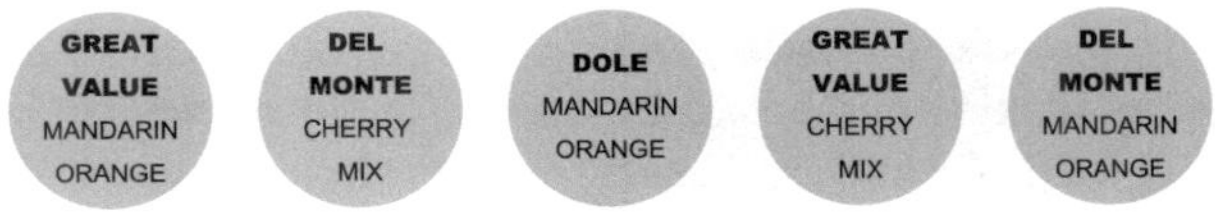

FIGURE 89 A position diagram showing the consumer preference ranking of five brands of fruit jelly. The data are based on *Consumer Report*, October, 2011.

It is important to note that our concept of basic semantic conventions is a historical one, not a logical one. That is, given the overall semantic relation $\Rightarrow$ supported by a representation system, the division of $\Rightarrow$ into $\rightharpoonup$ and $\rightharpoonup_D$ may not be logically unique. Even when a sub-relation $\rightharpoonup$ of $\Rightarrow$ can serve as the basis from which the rest of $\Rightarrow$ can be derived, there could be an alternative sub-relation $\rightharpoonup'$ that can serve as the basis equally well. In such cases, there is no logical way of determining which of $\rightharpoonup$ or $\rightharpoonup'$ represents the basic semantic conventions of the system.

Consider, for example, the simple position diagram in Figure 89. The basic semantic conventions associated with this system of position diagrams can be any of the following:

(37) If a circle with name X is to the left of a circle with a name Y, it means that the preference for the brand with name X is higher than that for the brand with name Y.

(38) If a circle with name X is to the right of a circle with name Y, it means that the preference for the brand with name X is lower than that for the brand with name Y.

(39) If a circle with name X is in the n-th position from the left, it means that the preference for the brand with name X is the n-th highest of them all.

(40) If a circle with name X is in the n-th position from the right, it means that the preference for the brand with name X is the n-th lowest of them all.

Assuming suitable syntactic and spatial constraints, these meaning relations can be derived from one another. Thus, when each is combined with the respective derivative meaning relation, they yield exactly the same overall meaning relation $\Rightarrow$.[24] Thus, logical analysis alone does not tell us which of (37)–(40) represents the system's basic conventions (and hence does not tell us which part of the meaning relation is a derived one, either).

---

relation defined by the abstraction tracking relation is the union of $\rightharpoonup$ and $\rightharpoonup_D$, and hence it is identical to $\Rightarrow$.

[24] I am grateful to Makoto Kanazawa for pointing this out to me, with an example similar to that above.

Generally, the issue of which part of $\Rightarrow$ is basic must be settled as a historical matter, by looking into how the system was introduced and inherited by the relevant group of people.[25] The notion of basic semantic conventions is a historical one in this sense.

5.2.2 Application to Specific Cases

We will now apply our central notion of abstraction tracking to account for some of the specific cases of derivative meaning observed in Section 5.1. Examples are selected, mainly for the reasons of their promise of revealing diverse patterns in which additional meaning relations are derived from relatively small sets of basic semantic conventions.

The Case of Line Graphs

Remember that one of the major functions of line graphs is to reveal differences in trends through differences in the orientations of lines (Kosslyn 1994). For example, the following instance of derivative meaning is supported by the system of line graphs cited in Subsection 5.1.2:

(41) If all lines but one line are slanted right-down, it means that all groups of people but one have lower parafabuloid levels at ages 65 and up than at ages under 65.

Let us call the source type involved in this instance σ_1 and call the target type θ_1. Thus:

(σ_1) All lines but one line are slanted right-down.

(θ_1) All groups of people but one have lower parafabuloid levels at ages 65 and up than at ages under 65.

We want to see how the instance $\sigma_1 \rightharpoonup_D \theta_1$ of derivative meaning holds in the present system of line graphs.

One illuminating way of analyzing this case of derivative meaning would be to start with the following, more elementary cases of derivative meaning:

(42) If a line tagged with sex S and income I is slanted right-down, it means that the parafabuloid level of people with sex S and income I is lower at ages 65 and up than at ages under 65.

(43) If a line tagged with sex S and income I is slanted right-up, it means that the parafabuloid level of people with sex S and income I is higher at ages 65 and up than at age under 65.

[25] Perhaps, different subsets of the group use a different one of (37)–(40) as their set of basic semantic conventions. In such cases, there are really four different systems of representations used by each subset of the group. Serious miscommunication is unlikely, however, as the overall meaning relation is the same between these systems.

(44) If a line tagged with sex S and income I is horizontal, it means that the parafabuloid level of people with sex S and income I is the same at ages under 65 and at ages 65 and up.

The analyses of these instances of derivative meaning are just routine, drawing on basic cases of the abstraction tracking relation. For the sake of non-redundancy, we will skip the analyses and simply assume that the validity of (42)–(44) can be derived from the system's basic semantic conventions.

We can then use these instances of derivative meaning to account for a more abstract case (41) of derivative meaning. Let us adopt the following abbreviations for ease of exposition:

R-dn(S,I)	=	a line tagged with sex S and income I is slanted right-down (source type)
R-up(S,I)	=	a line tagged with sex S and income I is slanted right-up (source type)
Level(S,I)	=	a line tagged with sex S and income I is horizontal (source type)
Decr(S,I)	=	the parafabuloid level of the people with sex S and income I is lower at ages 65 and up than at ages under 65 (target type)
Incr(S,I)	=	the parafabuloid level of the people with sex S and income I is higher at ages 65 and up than at ages under 65 (target type)
Same(S,I)	=	the parafabuloid level of the people with sex S and income I is the same at ages under 65 and at ages 65 and up (target type)

With these abbreviations, we can rewrite the cases of derivative meaning (42)–(44) in the following way:

(45) R-dn$(S,I) \rightharpoonup_D$ Decr(S,I)

(46) R-up$(S,I) \rightharpoonup_D$ Incr(S,I)

(47) Level$(S,I) \rightharpoonup_D$ Same(S,I)

It is not difficult to see that the iterative case of abstraction tracking in Figure 90 holds, mediated by these instances of derivative meaning.

In our analysis of the second case of derivative meaning of round-robin tables (See Appendix B), we noted the fact that a case of abstraction tracking *mediated by derivative meaning relation* (as opposed to basic semantic conventions) is sufficient for establishing a valid informational relation. Thus, the case of abstraction tracking in Figure 90 establishes a valid informational

$$\theta_1 \qquad \bowtie \qquad \left\{\begin{array}{l}
\left\{\begin{array}{ll}\text{Incr(Female, \$0–24,999),} & \text{Decr(Female, \$25,000+),}\\ \text{Decr(Male, \$0–24,999),} & \text{Decr(Male, \$25,000+)}\end{array}\right\},\\
\left\{\begin{array}{ll}\text{Same(Female, \$0–24,999),} & \text{Decr(Female, \$25,000+),}\\ \text{Decr(Male, \$0–24,999),} & \text{Decr(Male, \$25,000+)}\end{array}\right\},\\
\left\{\begin{array}{ll}\text{Decr(Female, \$0–24,999),} & \text{Incr(Female, \$25,000+),}\\ \text{Decr(Male, \$0–24,999),} & \text{Decr(Male, \$25,000+)}\end{array}\right\},\\
\left\{\begin{array}{ll}\text{Decr(Female, \$0–24,999),} & \text{Same(Female, \$25,000+),}\\ \text{Decr(Male, \$0–24,999),} & \text{Decr(Male, \$25,000+)}\end{array}\right\},\\
\left\{\begin{array}{ll}\text{Decr(Female, \$0–24,999),} & \text{Decr(Female, \$25,000+),}\\ \text{Incr(Male, \$0–24,999),} & \text{Decr(Male, \$25,000+)}\end{array}\right\},\\
\left\{\begin{array}{ll}\text{Decr(Female, \$0–24,999),} & \text{Decr(Female, \$25,000+),}\\ \text{Same(Male, \$0–24,999),} & \text{Decr(Male, \$25,000+)}\end{array}\right\},\\
\left\{\begin{array}{ll}\text{Decr(Female, \$0–24,999),} & \text{Decr(Female, \$25,000+),}\\ \text{Decr(Male, \$0–24,999),} & \text{Incr(Male, \$25,000+)}\end{array}\right\},\\
\left\{\begin{array}{ll}\text{Decr(Female, \$0–24,999),} & \text{Decr(Female, \$25,000+),}\\ \text{Decr(Male, \$0–24,999),} & \text{Same(Male, \$25,000+)}\end{array}\right\}
\end{array}\right\}$$

$$\upharpoonleft_{1\text{-}1}\upharpoonright$$

$$\sigma_1 \qquad \bowtie \qquad \left\{\begin{array}{l}
\left\{\begin{array}{ll}\text{R-up(Female, \$0–24,999),} & \text{R-dn(Female, \$25,000+),}\\ \text{R-dn(Male, \$0–24,999),} & \text{R-dn(Male, \$25,000+)}\end{array}\right\},\\
\left\{\begin{array}{ll}\text{Level(Female, \$0–24,999),} & \text{R-dn(Female, \$25,000+),}\\ \text{R-dn(Male, \$0–24,999),} & \text{R-dn(Male, \$25,000+)}\end{array}\right\},\\
\left\{\begin{array}{ll}\text{R-dn(Female, \$0–24,999),} & \text{R-up(Female, \$25,000+),}\\ \text{R-dn(Male, \$0–24,999),} & \text{R-dn(Male, \$25,000+)}\end{array}\right\},\\
\left\{\begin{array}{ll}\text{R-dn(Female, \$0–24,999),} & \text{Level(Female, \$25,000+),}\\ \text{R-dn(Male, \$0–24,999),} & \text{R-dn(Male, \$25,000+)}\end{array}\right\},\\
\left\{\begin{array}{ll}\text{R-dn(Female, \$0–24,999),} & \text{R-dn(Female, \$25,000+),}\\ \text{R-up(Male, \$0–24,999),} & \text{R-dn(Male, \$25,000+)}\end{array}\right\},\\
\left\{\begin{array}{ll}\text{R-dn(Female, \$0–24,999),} & \text{R-dn(Female, \$25,000+),}\\ \text{Level(Male, \$0–24,999),} & \text{R-dn(Male, \$25,000+)}\end{array}\right\},\\
\left\{\begin{array}{ll}\text{R-dn(Female, \$0–24,999),} & \text{R-dn(Female, \$25,000+),}\\ \text{R-dn(Male, \$0–24,999),} & \text{R-up(Male, \$25,000+)}\end{array}\right\},\\
\left\{\begin{array}{ll}\text{R-dn(Female, \$0–24,999),} & \text{R-dn(Female, \$25,000+),}\\ \text{R-dn(Male, \$0–24,999),} & \text{Level(Male, \$25,000+)}\end{array}\right\}
\end{array}\right\}$$

FIGURE 90 Iterative tracking of abstraction in the system of line-graphs, which accounts for the valid informational relation from source type σ_1 to target types θ_1.

relation from source type σ_1 to θ_1, accounting for the logical origin of the instance (41) of derivative meaning relation.

One interesting observation we can make from this analysis is that (i) certain cases of derivative meanings combine themselves to create an *indirect scale* with which we can measure objects in a graphic to measure objects in the represented situation, and that (ii) this pair of scales generates further instances of derivative meaning, where a comparison based on one scale indicates a comparison based on the other scale. In the present example, instances (45)–(47) of derivative meaning combine themselves to create a nominal scale (right-up/horizontal/right-down) on the lines in a graph that we can use as a substitute for another nominal scale (increased/same/decreased) on the parafabuloid levels of people at two different ranges of age. This is phase (i) above. Then, comparisons such as different/same orientations defined on one scale are informationally related to comparisons such as different/same tendencies defined on the other scale. This is phase (ii).

In fact, this (i)–(ii) pattern of multiplication of derivative meaning was encountered earlier, when we analyzed the case of round-robin tables. There, the instances of derivative meaning of the form $\bigcirc_n(X\text{-row}) \rightharpoonup_D W_n(x)$ (Subsection 5.2.1) combine themselves to create an absolute scale (number of white circles in table rows) that we can use as a substitute for another absolute scale (number of games won by teams). These linked scales then establish a valid informational relation from comparative properties (such as greater/same/smaller number of white circles) to other comparative properties (such as greater/same/smaller number of games won). Thus, (i)–(ii) seems to be a frequent, and therefore important, pattern that accounts for the proliferation of a derivative meaning relation based on a relatively small set of basic semantic conventions.

The Case of Node-Edge Graphs

The pattern (i)–(ii) is also present in the case of the London Underground maps discussed in Subsection 5.1.6, but this time it comes with the suggestion of another important pattern of meaning multiplication. There is an informational relation from the number of station icons directly connected to a particular station icon to the number of stations directly connected to a particular station. That is, a derivative meaning relation holds from source type σ_2 to target type θ_2 for every natural number n and every station x in the tube system:

(σ_2) n station icons are directly connected to a station icon for station x.

(θ_2) n stations are directly connected to station x.

The analysis of this instance of derivative meaning follows the usual path, which we will again skip to avoid redundancy.

Note that the relation $\sigma_2 \rightharpoonup_D \theta_2$ establishes linked absolute scales with which we can measure the number of stations directly connected to a station by the number of station icons directly connected to a station icon. These linked scales then entail further instances of derivative meaning in which greater/equal/fewer numbers of directly connected station icons indicate greater/equal/fewer numbers of directly connected stations. We are seeing the pattern (i)–(ii) here again.

Now, in Subsection 5.1.6, we also noted that a concentration of edges to a station icon indicates the extent to which the represented station is a hub for the entire tube system. The linked scales established by $\sigma_2 \rightharpoonup_D \theta_2$ above clearly contribute to this indication of "hub-ness," since the more stations icons are directly connected to a station icon, the more stations are directly connected to the represented station, and hence to a greater extent is the station a hub of the tube system. Of course, the categorization to hub-ness is usually a more complex matter than just measuring the number of directly connected stations. Other scales, such as the *ranking* of stations by directly connected stations or the number of different *kinds* of connected tube lines, can also contribute to the criterion. Although detailed analysis is beyond the scope of this book, nothing seems to prevent these scales from having indirect scales on the graphic side, based on the properties of station icons and edges connecting them. Combined, this set of indirect scales can generate a fine-grained scale for hub-ness, which can be deployed by a viewer with appropriate experience and expertise.

This example demonstrates another important pattern in which a graphical system obtains enriched semantic coverage through meaning derivation: (i′) creation of *multiple* pairs of linked scales, and (ii′) integration of these pairs into a new, multivariate pair of linked scales. Feature (ii′) is important insomuch as the represented multivariate scale is important for one's purpose of using a graphical system. For example, the degree of hub-ness is generally an important scale for a station, for the purpose of trip planning, as well as for the analysis and design of a train traffic system.

The Case of Scatter Plots

The degree of linear correlation of two numerical variables is fundamental to many purposes in data analysis, and the system of scatter plots cited in Subsection 5.1.1 provides an indirect scale for this degree through the meaning derivation pattern (i′)–(ii′). Given a set l of points that makes up a line on a plane, the distance of a dot from l derivatively means the deviation of a data point from a linear equation. (As this is again a routine to analyze this instance of derivative meaning, so we will skip it here.) On the basis of these linked scales, another pair of linked scales for the average distance of all dots from l and hence for the average deviation of all data points from the equation is defined. This pair in turn defines still another pair of scales for the angle of the line that maximally approximate the distribution of dots on a plot and hence for the coefficient of the linear equation that maximally approximates the data distribution. Combined, these pairs of linked scales determine an indirect scale for the linear correlation of the relevant variables.

Similar analyses in terms of the pattern (i′)–(ii′) apply to account for the functions of scatter plots to express the degree of quadratic curvlinear correlation, the degree of exponential curvlinear correlation, and the degree of outlier-ness, although we will not work out the actual analyses here.

The Case of Contours Maps

The capacity of contour maps to express the degree to which a terrain is shaped like a valley, a ridge, or a cirque (Section 5.1.4) accepts the explanation based on the pattern (i′)–(ii′), but this only means that contour maps can express the degree in which a terrain fits various *shape* categories, which should be distinguished from geographical categories. Graded categorization into the latter categories typically involves not only the shape of a terrain, but the cause for it to take that shape. To be a cirque, for example, the cirque shape must be caused by glacial erosion. The cause of the shape of a terrain is definitely outside the expressive coverage of standard contour maps, even with their meaning derivation properties. Thus, contour maps generally cannot express the degree to which a terrain *is* a valley, a ridge, or a cirque. Derivative meaning multiplies itself to extend the system's expressive coverage, but never to cover categories that involve something that cannot be expressed by two-dimensional, planer patterns.

5.3 Questions and Responses

5.3.1 Question: Differences from Secondary Notations?

According to Petre and Green (1992), the users of a graphical representation system frequently adopt semantic techniques that go beyond the scope of the system's basic semantic conventions. For example, skilled designers of electric circuit diagrams indicate the relatedness of circuit components by placing their symbols in adjacent positions in the diagram, although no such meaning relation is part of the basic semantic conventions of the circuit diagrams. Such skilled designers also avoid awkward distances between the symbols of related circuit components as they could miscue their unrelatedness.

Figure 91 is one of the circuit diagrams cited by Petre and Green (1992). Here, the fact that a set of components collectively makes a larger component of the circuit (such as counter components and buffer components) is indicated by the placement of their symbols in close proximity. Strictly speaking, however, the spatial positions of these symbols should be meaningless. According to basic semantic conventions, the only meaningful relation in an

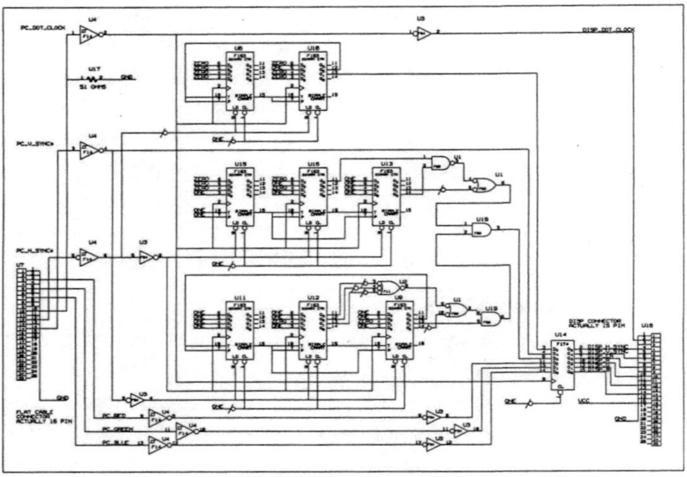

FIGURE 91 An electric circuit diagram following semantic strategies based on the secondary notation (Petre and Green 1992). Reprinted with permission.

electric circuit diagram is the connection of pairs of component symbols by actual lines that indicate the physical wiring of pairs of relevant components. The expression of their "relatedness" by spatial adjacency of symbols is based on an unofficial, "secondary" notation, "adjunct to" the basic notation (p. 67).

Meaning by a secondary notation, therefore, goes beyond what is directly justified by a system's basic semantic conventions, and in this respect, it is similar to derivative meaning. Do they make up the same class of meaning relations, or is one a species of the other?

Response

The difference between derivative meaning and meaning based on secondary notation becomes clear when we consider their logical relation to the system's basic semantic conventions.

As its name indicates, a derivative meaning relation is *derivable* from basic semantic conventions on the basis of the system's abstraction tracking properties. Thus, given basic semantic conventions, the derivative meaning relation is unavoidable. In contrast, the extra informational relation based on secondary notation is an optional addition, or an "adjunct," to the system's semantic conventions, rather than something that necessarily follows from them.

Thus, for example, it is logically possible for the system of circuit diagrams with all its basic semantic conventions *not* to support the extra informational relation based on adjacency and remoteness of symbols. In fact, the system of circuit diagrams used by novices or unskilled designers can be considered to be such a simplistic, or "purebred" system, where the adjacency or remoteness of symbols means nothing (See Petre and Green (1992) for an example of a circuit diagram of this sort). Although such circuit diagrams are considered "awful" by those who are used to expecting the secondary notations, this is not because they fail to respect derivative meaning relations (which would be impossible), but because they fail to respect secondary notations.

Informational relations based on secondary notations are thus optional adjuncts to a system's basic semantic conventions, rather than an unavoidable consequence of them. The addition of a secondary notation to the primary system is a "manipulable variable," subject to "personal style and skill, and is not used well automatically" (Petre and Green 1992: p. 57). Precisely because of this characteristic, "secondary notation" is a powerful concept that accounts for an important kind of miscommunication caused by graphics. It also accounts for the source of the functional difference in diagrams drawn by experts and novices (Petre 1995: p. 43). The concept should however be strictly distinguished from the notion of derivative meaning, which identifies the utilization of valid informational relations necessarily generated from the given choice of basic semantic conventions.

5.3.2 Question: Difference from Graphical Implicatures?

What about the notion of graphical implicatures introduced by Marks and Reiter (1990)? This also captures meaning relations that go beyond the basic semantic conventions in a graphical representation system.

Response

Let us examine the examples of graphical implicatures given by Marks and Reiter (1990). Compare the two diagrams, given in Figure 92-a and 92-b, that represent an identical computer disk system.

According to Marks and Reiter (1990), Figure 92-a carries at least three (unwanted) implicatures absent from Figure 92-b. First, the pen-width used for one of the channel-facility queue symbols (in the central part of the diagram) differs from the pen-width used for the other queue symbols. This fact implicates that the particular channel-facility queue denoted by this symbol is somehow different from the other queues because otherwise the diagram designer would simply have used the same pen-width for it, as is the case with Figure 92-b. Second, the device queue symbols (in the upper part) are not exactly aligned horizontally. The implicature is that one of the device queues is uniquely different, for otherwise the designer would have arranged all the symbols in a line, as in Figure 92-b. Third, the way the disk symbols (in the lower part) are perceptually grouped into two "gestalts" implicates that there

(a)

(b)

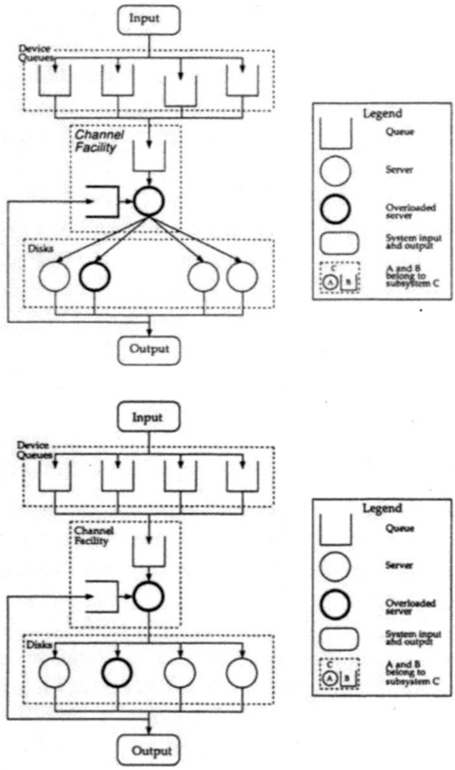

FIGURE 92 Network diagrams representing an identical computer disk system (Marks and Reiter 1990). Diagram (b) carries at least three (unwanted) graphical implicatures not carried by diagram (a). Reprinted with permission. Copyright © 1990 by Association for the Advancement of Artificial Intelligence.

is some semantic basis to this grouping, since otherwise the disk symbols would have been placed at equal distances, as in Figure 92-b.

Here, the inference take the following steps: (1) given a specific graphic, one finds something irregular in its appearance, namely, a violation of some standard of "preferred" look for graphics, (2) this finding prompts one to think about why the graphic must have been drawn that way, (3) one then hypothesizes the existence of a particular syntactic or semantic rule as the best explanation for this deviation from the standard, (4) in case a semantic rule is hypothesized, one re-interprets the graphic with this new rule, attributing new meaning to the feature identified in (1). Thus, in the case of Figure 92-a, the reader would attribute a special meaning to the pen-width of the particular queue symbol, to the irregular horizontal position of one of the device queue symbols, and to the grouping of the four disk symbols.

As the above formulation indicates, "graphical implicature" is a notion defined on *how we discover a meaning*. It is an *epistemological* notion in this sense. In contrast, "derivative meaning" is a notion defined on *how a meaning is originated*. The two concepts crosscut one another, so it is theoretically possible that derivative meaning is also a graphical implicature in a particular situation, if it is discovered through an inference for the best explanation (1)–(4) described above. Yet, neither the notion of graphical implicature exhausts derivative meaning (there are other ways derivative meaning is discovered in a given situation) nor is it limited to derivative meaning (the three examples of graphical implicatures cited earlier are all cases of secondary notations). As an epistemological concept, "graphical implicature" does not replace the semantic notion of derivative meaning.

5.3.3 Question: Relevance to Cognitive Potentials?

The notion of abstraction tracking captures a new, distinctive species of meaning relation typical to graphical representation systems. So, your analysis of derivative meaning might be of significant semiotic interest.

But what about its cognitive significance? As you have abundantly documented, researchers were well aware of what you call derivative meanings in graphics. Although their exact logical origin has never been made an issue of, some pioneering psychological studies have revealed the way they affect our comprehension of graphical representations. Under this circumstance, does your analysis add anything to our knowledge about their cognitive potentials?

Response

As our examples in Section 5.1 have already suggested, it is a more or less common assumption in graph-comprehension research that there is a clearly distinguishable class of reading operations that directly interpret the "global" or "higher-order" perceptual properties of graphical representations. The

"Graph Difficulty Principle" proposed by Pinker (1990) states that graphs are easy to comprehend to the extent that direct translation routines on such global patterns can be utilized (p. 108). Lohse (1993) characterizes such a routine "perceptual inference made directly" (p. 382), and adopts it as an important parameter in his computational simulation of the way people comprehend visual representations. Gattis and Holyoak (1996) even investigated how persistent a direct translation routine was over a significant change in representation schema.

The process of directly interpreting a global pattern then has been contrasted to the process that interprets "local" or "elementary" features of graphics. Bertin (1967/1973) started the tradition of distinguishing different kinds of questions to be asked about data expressed in statistical charts, especially (a) "global-search" or "higher-level" questions demanding the identification of the overall trends or the comparison of multiple data values, and (b) "local-search" or "lower-level" questions demanding the extraction of individual data values. Guthrie et al. (1993) and Ratwani et al. (2008) investigated how different kinds of questions prompted different reading strategies, including the direct translation routine.

The boundary between global reading and local reading, however, has never been studied explicitly. Pinker's theory (1990) offers a systematic account of the conditions for a reader to be engaged in direct translation operations, yet as a psychological theory, it is silent about the logical difference between the two classes of reading operations. Against this background, our analysis clarifies the contrast between the two classes of reading operations in terms of the meaning relation directly produced by historically defined basic semantic conventions and that derivable from it through an abstraction tracking structure. Our analysis therefore can serve as a useful conceptual footing to empirical investigation into the cognitive effects of global reading operations.

5.3.4 Question: Too Many Cases of Abstraction Tracking?

The definition of abstraction tracking is rather inclusive, resulting in numerous instances of derivative meaning that are unlikely to be put into use under any normal circumstances. For example, it is easy to see that σ_3 and θ_3 below stand in the abstraction tracking relation in the system of round-robin tables discussed in Subsection 5.1.7.

(σ_3) Either a white circle appears at both the intersection of a row headed by "CHN" and a column headed by "TUR" and at the intersection of a row headed by "USA" and a column headed by "KOR" or else there are more black circles in a column headed by "SRB" than in a column headed by "BRA."

	USA	CHN	KOR	BRA	TUR	SRB
USA	-	○	○	○	○	○
CHN	●	-	○	●	○	○
KOR	●	●	-	○	●	○
BRA	●	○	●	-	○	○
TUR	●	●	○	●	-	○
SRB	●	●	●	●	●	-

FIGURE 93 Round-robin tables reporting the results of a preliminary round of women's volleyball at the 2012 London Olympic games.

(θ_3) Either the China team has won against the Turkey team and the USA team has won against the South Korea team or else more teams have lost to the Serbia team than to the Brazil team.

This instance $\sigma_3 \rightharpoonup_D \theta_3$ of derivative meaning seems to be of no use under normal circumstances. Even when σ_3 holds of the round-robin table (see Figure 93 shown again), we normally do not look at the table so as to recognize σ_3 and read θ_3 off from it. The notion of abstraction tracking is too broad to exclude garbage of this sort. Perhaps, we need a more narrowly defined semantic notion to capture the truly functional aspect of a derivative meaning property.

Response

First, the instance $\sigma_3 \rightharpoonup_D \theta_3$ of derivative meaning is not garbage. True, σ_3 will not be the property that your perceptual system normally registers when you casually browse the table in Figure 93. However, *if* you do recognize σ_3 in the table, and *if* you have a reason to interpret σ_3, you will certainly interpret it to mean θ_3. This indicates that the instance $\sigma_3 \rightharpoonup_D \theta_3$ is *potentially functional.*

Second, one prominent use of a graphical system within the context of communication is in what may be called "open-ended communication." When someone presents a graphical representation to someone else, the presenter typically expects the viewer to obtain *whatever* information is expressed in his graphic depending on the viewer's informational needs and interests, and the viewer expects *whatever* information is expressed in the graphic to be true according to the presenter's beliefs. Thus, the information that enters into the viewer's belief system is fundamentally indeterminate, and be of any of the pieces of information expressed in the graphic. This is in stark contrast to our everyday verbal communication, where a speaker expects the hearer to obtain the *particular* piece of information expressed in his or her sentence, and the hearer expects *this* piece of information to be true according to the speaker's beliefs. The information that enters into the hearer's belief system is much more deterministic.[26]

For example, when someone presents a table such as that in Figure 93 to someone else, the presenter does not expect the viewer to obtain any particular piece of information expressed in it. The viewer may obtain information that the South Korea team won against the Brazil team (basic meaning), or that the Turkey team lost to three teams (derivative meaning), or that the USA team won against more teams than any other team did (again derivative meaning). The presenter's goal is fulfilled if the viewer reads off whatever information is relevant to his needs and interests. In fact, we can even think of a possible situation in which the viewer's informational needs are met when he or she reads off information θ_3 by interpreting σ_3.[27]

For the purpose of open-ended communication outlined above, the existence of a large class of potentially functional instances of derivative meaning is essential. It is these instances that support a high degree of open-ended-ness of communication with a graphical system. Our concept of abstraction tracking, by being inclusive, can do justice to this aspect of communication with graphics.

5.4 Chapter Takeaways

- Meaning derivation is a potential of a graphical representation system, where the meaning relation is expanded by derivation from the basic semantic conventions in the system.
- Derivative meaning relations are supported by many graphical systems, including various systems of statistical charts, maps, node-edge graphs, and iconic tables.
- A derivative meaning relation holds when a source type tracks a target type in abstraction, i.e., when the families of specific properties they abstract over correspond to one another according to the system's basic semantic conventions.
- Many instances of global reading or direct translations of perceptual patterns in a graphic are based in derivative meaning relations characterized in this way.

---

[26] See Lewis (1975/1985) for the model of verbal communication outlined here (pp. 7–10). Although an open-ended communication with a graphical representation does not fit in that particular model, it does fit in his general concept of communication based on a convention of truthfulness and trust.

[27] Imagine that your spouse and son are interested in the final result of the round-robin competition. Your spouse is a big fan of the Serbia team and interested in how it compares to the Brazil team in the number of teams they beat. Your son is a fan of the China and the USA teams and wishes them to win their final games, where they are respectively matched with the Turkey and the South Korean teams. You yourself are interested in whether either your spouse or son will be happy when they learn the final results of the competition.

- Derivative meaning is a distinctive species of meaning relation, to be conceptually distinguished from meaning based on secondary notation and graphical implicatures.

6

Extending the Analysis

We have now gone through four recurrent themes in modern discussions on the cognitive potentials of diagrams. The main finding thus far is then that some of the most important cognitive potentials of diagrams are rooted in semantic properties of diagrammatic systems that accept precise logical characterization.

Now, it is possible that the semantic properties thus extracted are responsible for *some other* cognitive potentials of diagrams than the ones that prompted their identification, and if this is the case, it will be an indication of the further utility of relevant concepts. In this chapter, we will demonstrate that this is indeed the case for the concept of consequence tracking. Specifically, we will consider three other cognitive potentials of diagrams that may be called *envisaging*, *aspect shifting*, and *transfer of spatial analysis*, and explain how the concept of consequence tracking might be applied to clarify the logical basis of these phenomena. This means that multiple cognitive processes can be triggered by consequence tracking properties of a representation system, and we will suggest that the rich cognitive potentials of so-called *law-encoding diagrams* (Cheng 1996, 1997, 2002) are partly attributable to this fact.

None of these extensions are final, however. Each theme represents a complex phenomenon that requires more detailed treatment than we can offer in this chapter. Nevertheless, the lines of analysis outlined here seem to offer hypotheses that are plausible enough to inspire more detailed and systematic research on these important themes in diagram research.

6.1 Envisaging

Back in Chapter 1, we considered the case where a simple position diagram is used to obtain a free ride. Here, we will consider slightly modified examples. We are given the position diagram in Figure 94, which expresses premise θ_1.

(θ_1) Mantle is better than Mays.

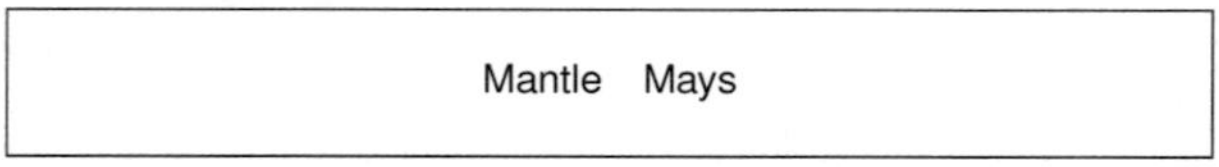

FIGURE 94 Position diagram expressing premise σ_1.

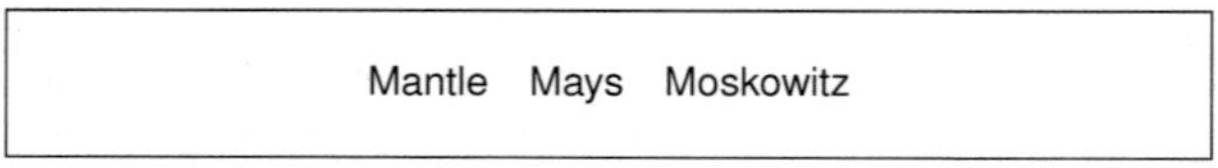

FIGURE 95 Position diagram expressing premises θ_1 and θ_2.

Along with this diagram, you are given the information θ_2 from a separate source, such as in speech or writing.

(θ_2) Mays is better than Moskowitz.

Then, we are asked, "Is Mantle better than Moskowitz?" One way of answering this question is to physically express θ_2 in the given diagram in Figure 94, by putting the name "Moskowitz" to the right of the name "Mays." This results in the diagram in Figure 95, which necessarily has the name "Mantle" to the left of "Moskowitz." You can then use this fact to read off information θ_3:

(θ_3) Mantle is better than Moskowitz.

Thus, the answer to our question turns out to be "Yes." This is a standard case of a free ride. So, the general analysis in Chapter 2 applies, and we can see the consequence tracking relation depicted in Figure 96 is responsible for this free ride.

Now, there is another important way of solving this problem. Instead of physically placing the name "Moskowitz" to the right of the name "Mays," we can *envisage* the placement of the name in that position. We can then "observe" that the name "Mantle" comes to the left of the name "Moskowitz" and interpret it to indicate information θ_3.

Although this second strategy is significantly different from the first strategy in the cognitive operations involved in it, it is important to understand that it depends on the same underlying semantic property as the first strategy does. Without the source constraint $\{\sigma_1, \sigma_2\} \vdash \sigma_3$ in Figure 96, envisaging the name "Moskowitz" to the right of the name "Mays" could fail to result in the name "Mantle" coming to the left of the name "Moskowitz." Without the indication relation $\sigma_3 \Rightarrow \theta_3$ in Figure 96, the result that "Mantle" is to the left of "Moskowitz" would not indicate information θ_3. Without the target $\{\theta_1, \theta_2\} \vdash \theta_3$ in Figure 96, the indicated information θ_3 would not be a consequence of premises θ_1 and θ_2. The inference based on envisaging is thus rooted in a consequence tracking property of the system of position diagrams, as the inference based on a free ride does.

$$
\left\{\begin{array}{cc} (\theta_1) & (\theta_2) \\ \text{Mantle is better} & \text{May is better than} \\ \text{than Mays,} & \text{Moskowitz} \end{array}\right\} \vdash \begin{array}{c} (\theta_3) \\ \text{Mantle is better} \\ \text{than Moskowitz} \end{array}
$$

$$
\Uparrow \qquad\qquad \Uparrow \qquad\qquad \Uparrow
$$

$$
\left\{\begin{array}{cc} (\sigma_1) & (\sigma_2) \\ \text{Mantle's name is to} & \text{Mays' name is to} \\ \text{the left of Mays'} & \text{the left of} \\ \text{name,} & \text{Moskowitz's name} \end{array}\right\} \vdash \begin{array}{c} (\sigma_3) \\ \text{Mantle's name is to} \\ \text{the left of} \\ \text{Moskowitz's name} \end{array}
$$

FIGURE 96 Consequence tracking property underlying a free ride and an envisaging on position diagrams.

Here, we are only concerned with the logical requirements for an inference based on envisaging to be valid, rather than the cognitive processes involved in it. Regarding the latter issue, there are two alternative approaches. The first is the internalist approach, where envisaging is considered to be an operation applied to a mental copy of the relevant diagram. One implication of this view is that people somehow internalize the relevant system of diagrams into their inferential engines, reflecting the structural constraints on diagrams necessary for envisaging. The second is the externalist approach, where envisaging is considered to be an operation extended to an external diagram with the help of the visual indexing abilities of humans (Pylyshyn 1989; Pylyshyn and Storm 1988; Pylyshyn 2003). Shimojima and Katagiri (2013) reported on eye-tracking data that clearly show the existence of envisaging operations on diagrams as postulated by diagram researchers. Although their experiments were based on the externalist approach to envisaging, their results were also compatible with the internalist approach. We will discuss this work in some detail at the end of this section.

Let us consider another example of envisaging, taken from an experiment conducted by Bauer and Johnson-Laird (1993). Suppose you are given the following premises in writing:

(θ_4) Raphael is in Tacoma or Julia is in Atlanta, or both.

(θ_5) Julia is in Seattle or Paul is Philadelphia, or both.

The task is to consider various possible situations that satisfy these premises, and draw a conclusion that is common to all these situations. This is a fairly difficult problem if we try to solve it by just looking at these premises in written form.

Things will change, however, if we are also supplied with "quasi-circuit diagrams" of the kind illustrated in Figure 97. In this system of diagrams, a shape carrying a person's name can only be placed in a slot with a corre-

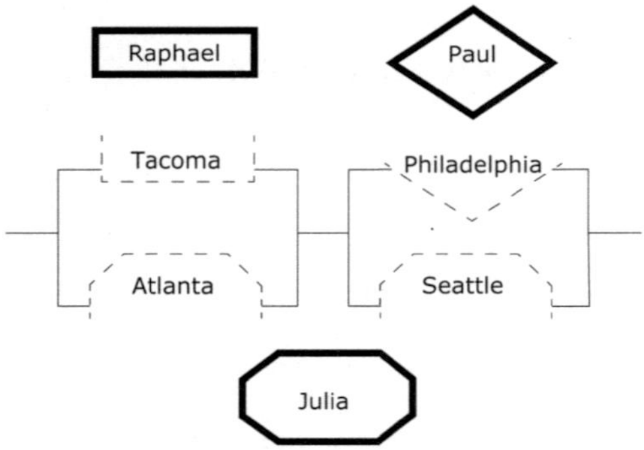

FIGURE 97 A sample quasi-circuit diagram used in an experiment by Bauer and Johnson-Laird (1993). The original picture appears on page 7 and is reprinted with permission of SAGE Publications. Copyright © 1993 by SAGE Publications.

sponding shape, and if so placed, it indicates that the relevant person is in the city denoted by the slot. Thus, when our two premises are expressed in the system, they amount to the following conditions:

(σ_4) A shape labeled "Raphael" occupies a slot labeled "Tacoma" or a shape labeled "Julia" occupies a slot labeled "Atlanta," or both.

(σ_5) A shape labeled "Julia" occupies a slot labeled "Seattle" or a shape labeled "Paul" occupies a slot labeled "Philadelphia," or both.

The conjunction of these conditions in turn is equivalent to the following condition:

(σ_6) There is a path, consisting only of lines and occupied slots, from one side of the diagram to the other.

Therefore, we can express various situations that satisfy the given premises θ_4 and θ_5 by placing the three shapes carrying people's names in the given slots in a way that satisfies condition σ_6.

However, we do not have to physically draw these shapes in the city slots to do this, but can simply envisage various possibilities of such placements. For example, we can envisage Raphael's shape in Tacoma's slot and Julia's shape in Seattle's slot, and "observe" that such a placement of shapes satisfies condition σ_6. Or we can envisage Julia's shape in Atlanta's slot and Paul's shape in Philadelphia's slot, and "observe" that the condition σ_6 is satisfied. This way, we can consider various possibilities that satisfies σ_6, which correspond to various possible situations that satisfy premises θ_4 and θ_5. In their experiment, Bauer and Johnson-Laird observed a significant improvement in performance (percent correct and latency) by participants supplied with these quasi-circuit diagrams, and they attributed the cause of the improvement to

{ (θ_6) Julia is in Atlanta, (θ_7) Paul is in Philadelphia }	$\vdash$			(θ_4) $\wedge$ (θ_5) Raphael is in Tacoma or Julia is in Atlanta, or both, and Julia is in Seattle or Paul is in Philadelphia, or both
$\Uparrow$				$\Uparrow$
{ (σ_7) Julia's shape occupies Atlanta's slot, (σ_7) Paul's shape occupies Philadelphia's slot }	$\vdash$	(σ_6) There is a path, consisting only of lines and occupied slots, from one side of the diagram to the other	$\equiv$	(σ_4) $\wedge$ (σ_5) Raphael's shape occupies Tacoma's slot or Julia's shape occupies Atlanta's slot or both, and Julia's shape occupies Seattle's slot or Paul's shape occupies Philadelphia's slot, or both

FIGURE 98 Consequence tracking property underlying an envisaging on quasi-circuit diagrams. The symbol $\equiv$ stands for equivalence under the constraints on the domain of quasi-circuit diagrams.

the existence of an envisaging process based on these diagrams.

Note that the envisaging process in this example again depends on the consequence tracking property of the relevant system of diagrams. Figure 98 shows the consequence tracking property underlying one of the envisioning processes mentioned above. If it were not for the bottom constraint $\{\sigma_7, \sigma_8\} \vdash \sigma_6$ ($\equiv \sigma_4 \wedge \sigma_5$), envisaging Raphael's shape in Tacoma's slot and Julia's shape in Seattle's slot would not allow one to verify that σ_6 was satisfied. Or if the tracked constraint $\{\theta_6, \theta_7\} \vdash \theta_4 \wedge \theta_5$ were not actually a constraint, the verification that the particular placement σ_7 and σ_8 of shapes satisfied condition σ_6 ($\equiv \sigma_4 \wedge \sigma_5$) would provide no guarantee that the corresponding assumptions θ_6 and θ_7 would satisfy the premises θ_4 and θ_5.[28]

These two examples demonstrate that the consequence tracking properties of a diagrammatic system play a crucial role in the envisioning process on

[28]Note that the identification of source type σ_6 is crucial to the effective solution of the present problem. Source type σ_6 is equivalent to source type $\sigma_4 \wedge \sigma_5$ indicating the crucial condition $\theta_4 \wedge \theta_5$, yet σ_6 is significantly easier to recognize than $\sigma_4 \wedge \sigma_5$. In fact, the existence of such an "equivalent cue" is the whole point of using circuit-diagrams in expressing conjunctions of inclusive disjunctions. Although our present focus is on the envisaging phenomenon, such an equivalent cue in the use of diagrams seems to be a phenomenon worth investigating separately.

diagrams. It helps us to account for the fact that an envisaging process on diagrams can afford efficient and accurate inference about the situations that are represented. In our view, this account applies to many cases of envisaging studied in the previous research, including the cases of hinge diagrams (Schwartz 1995), mechanical behavior-hypothesis diagrams (Narayanan et al., 1995), rod-joint diagrams (Sloman 1971), geometry diagrams (Giaquinto 2007), and topographical maps (Chandrasekaran et al. 2011).

There are, however, other cases of envisaging that our present concept of consequence tracking does not cover. Sometimes, when the movements or changes in diagrams are envisaged, we not only reason about the results of the movements or changes, but the movements and changes themselves. For example, when one uses the diagram in Figure 99-a to figure out the most efficient way of moving a plank to bridge a ditch, the focus is on the movement of the rectangle representing the plank and the corresponding movement of the real plank (Sloman 1971). Another example is when one uses the pulley diagram in Figure 99-b to figure out the movements of pulleys caused by pulling on the rope in the downward direction (Hegarty 1992/1995).

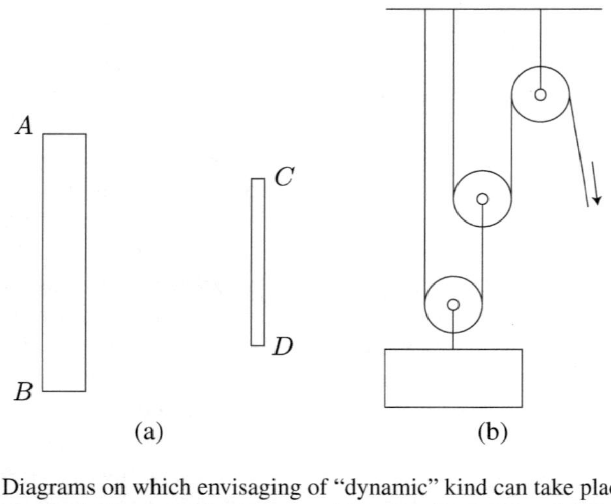

FIGURE 99 Diagrams on which envisaging of "dynamic" kind can take place. Try to figure out: (a) what is the most efficient way of moving plank *CD* to bridge ditch *AB* (Sloman 1971), and (b) what movements of pulleys will be caused by pulling the rope in the downward direction (Hegarty 1992/1995). Diagrams (a) is reprinted with permission of Elesevier.

Handling these cases would require our model to cover constraints governing movements or changes and their initial and final conditions. Since our model currently only covers constraints governing static objects or situations, this requires a significant extension of our conceptual framework. Although we do not foresee any unsurpassable obstacles to such an extension, it does require careful treatment and must be left for future developments.

Aside: Externalist account for envisaging based on visual indexing

Shimojima and Katagiri (2013) proposed an externalist account that is applicable to non-dynamic cases of envisaging. Behind their account is the theory of visual indexing that has recently received substantial empirical supports (e.g., Pylyshyn 1989; Pylyshyn and Storm 1988; Ballard et al. 1997; Pylyshyn 2003; Spivey et al. 2004). According to this theory, we are endowed with maintaining a small number of visual indexes (typically, four or five) at different objects or locations in the visual scene. The main functions of these indexes are to let us (1) keep track of multiple objects in the visual scene, (2) return focal attention to any indexed object or location without a fresh search, and (3) associate with an indexed object or location some task-relevant information stored in internal memory.

With the background of this theory, Shimojima and Katagiri (2013) analyze an envisaging operation on a diagram in the following way. Whenever one wants to add a piece of information to a diagram, one could actually place an object with an appropriate visual property in an appropriate location of the diagram, or instead, they could place a *visual index* at the appropriate location, associating with it some task-relevant information stored in internal memory. In the first case, one could perceive what spatial relation or pattern is produced as a result of the new object being added and use this observation to draw a conclusion of the inference. This is the usual process of a free ride, based on an actual drawing. In the second case, one can also observe the resulting spatial relation or pattern, but this time, the one that was produced by adding the visual index.

For example, the newly indexed location may have a certain spatial relation (say, ABOVE) with two other objects in the diagram, and it may produce a particular shape (say, a triangle) in combination with these objects. This observation can then be used to draw a conclusion for the inferential task, and when this happens, it is a case of free ride, based on the spatial constraint on diagrams. Or in the case of a position diagram, one may place a visual index to the right of Mays' name in Figure 94. Then, due to spatial constraints on locations and objects on a two-dimensional plane, Mantle's name is necessarily located to the left of this indexed location. This spatial relation can be observed through a usual perceptual process, as the releta are an external object (a name) and a location on an external plane. The observation can then be used to draw a conclusion for the inferential task. The only difference from the usual process of a free ride is that the observed spatial relation or pattern is partly formed by an indexed location, whose visual feature is stored not *in situ* but in the internal memory. This model thus explains the free-ride-like ef-

TROUBLE SPOTS	Chrysler Town & Country						Dodge Caliber					
	'06	'07	'08	'09	'10	'11	'06	'07	'08	'09	'10	'11
Engine, major	○	○	○	○	○	○		○	○	○	○	★
Engine, minor	◕	◓	◔	○	○	○		◔	○	○	○	★
Engine, cooling	●	◓	◓	○	○	○		◓	○	◔	○	★
Trans., major	○	◔	◔	○	○	○		◓	◔	○	○	★
Trans., minor	○	◔	◓	◓	◔	◔		◔	○	○	○	★
Drive system	◔	◓	◔	◔	○	○		◕	○	○	○	★
Fuel system	◕	◕	◓	◔	○	○		◕	◕	○	◔	★
Electrical	◕	◓	◓	◔	◔	○		◕	◔	○	○	★
Climate system	●	●	●	●	◔	◓		●	◔	◔	◔	★
Suspension	●	●	●	◓	○	○		●	●	◓	○	★
Brakes	●	●	●	●	●	◔		◕	◓	◕	●	★
Paint/trim	◕	●	◕	◔	○	○		●	◕	◔	◔	★
Body integrity	◕	●	●	●	◕	◓		●	●	◕	◓	★
Body hardware	●	●	●	●	◕	◕		◓	○	○	◓	★
Power equipment	◕	◕	●	◕	◕	●		◓	◓	◓	◓	★
Audio system	◓	◓	●	◓	◓	◓		◓	○	○	◔	★

TROUBLE SPOTS	Dodge Challenger						Dodge Charger					
	'06	'07	'08	'09	'10	'11	'06	'07	'08	'09	'10	'11
Engine, major			★	○	○	★	○	○	◔	○	★	★
Engine, minor			★	○	○	★	○	◔	○	○	★	★
Engine, cooling			★	○	○	★	◕	◕	○	◔	★	★
Trans., major			★	○	○	★	○	○	○	○	★	★
Trans., minor			★	◕	○	★	●	●	◔	◓	★	★
Drive system			★	○	○	★	◔	◕	◔	◔	★	★
Fuel system			★	◔	○	★	◔	◕	○	○	★	★
Electrical			★	○	○	★	◔	◔	○	○	★	★
Climate system			★	○	◓	★	◓	◕	○	◔	★	★
Suspension			★	○	○	★	●	●	●	◕	★	★
Brakes			★	◓	◕	★	◕	●	◓	◓	★	★
Paint/trim			★	◓	◔	★	◕	●	◔	◓	★	★
Body integrity			★	◔	◓	★	●	◓	◔	◓	★	★
Body hardware			★	◔	◓	★	◓	◓	○	◓	★	★
Power equipment			★	◓	◔	★	◓	◓	◓	◕	★	★
Audio system			★	◓	◔	★	◕	◕	◓	◔	★	★

FIGURE 100 Table simulating a function of "frequency of repairs" tables periodically published in *Consumer Report*. The data are based on the 2012-April issue.

fect of envisaging, on the basis of operations on external diagram, as opposed to operations on internalized versions of external diagrams.

6.2 Transfer of Spatial Analysis

The U.S.-based magazine *Consumer Report* periodically publishes "frequency of repairs" tables for automobiles. The table in Figure 100 conveys the feel of those popular tables, although it is grossly simpler than the actual ones. Each of the four horizontally arranged mini-tables represents an automobile brand such as Chrysler's Town & Country and Dodge's Caliber, where the six columns in each mini-table represent models of years 2006 through 2011. The headers of rows list various sorts of repairs classified by the parts repaired (e.g., engines, transmissions, and drive systems) and the scale of repairs (major or minor). The kinds of symbols appearing at the intersections of rows and columns indicate the frequency of repairs reported for the relevant automobile model with respect to the relevant sorts of repairs: a white circle in the center indicates relatively few occurrences of repairs, a full black circle indicates significantly more repairs than average, and other mixed-color circles fall in between, with a larger ratio of the black area indicating a higher frequency of repairs. A blank column indicates that the automobile brand has no model in that year, while a column of star symbols means that data are insufficient for that year's model.

Assuming the repairs on the power train are regularly listed in the upper part of the table, one prominent trend shown in this table is that Chrysler's Town & Country has a high frequency of repairs outside the power train. We can read this information off from the fact that the mini-table labeled "Chrysler Town & Country" is blackish in the lower part. This is clearly a case of the derivative meaning discussed in Chapter 5, where "being blackish in the lower part" plays the role of abstract source type σ in our analysis and "having a high frequency of repairs outside the power train" plays the role of abstract target type θ.

Now, after reading off information θ that Chrysler's Town & Country has a high frequency of repairs outside the power train, one may naturally wonder what exactly is wrong with its non-power-train aspect—namely, what more specific facts about this automobile brand make up this rather unwelcome trend. One would be especially keen to make such queries if one owned a car of this brand or were considering purchasing one. One natural thing to do to answer this question is to look more into the graphic at hand, namely, to look into the lower part of the mini-table labeled "Chrysler Town & Country" and inspect exactly what makes that part blackish. One may find it is partly because the row labeled "Climate system" has four full black circles, partly because the row labeled "Suspension" has three full black circles, and so on. Further, these facts mean, again by derivative meaning, that the climate system had significantly more frequent repairs in four models of this brand, that the suspension had significantly more frequent repairs in five models, and so on. Therefore, one would take these to be the specific facts that make up the general trend that Chrysler's Town & Country had a high frequency of repairs outside the power trains.

All this process of data analysis looks too natural to raise anybody's eyebrows. The reality is that an important substitution of tasks is taking place in this process. Remember that the original question one started with was what exactly was wrong with the non-power-train aspects of Chrysler's Town & Country. However, the question that one directly investigated was different, and was concerned with why exactly the lower part of the box labeled "Chrysler Town & Country" was blackish. Accordingly, one investigated individual rows that made up the relevant part of the box, and found several specific properties of these rows that made that part blackish. Using the derivative meaning relation, one then translated these properties of rows into specific properties of the automobile brand, and identified these properties as the factors that made up the general trend in question. The original task of analyzing the high frequency of repairs on a trouble area of an automobile brand is replaced by a *spatial-analysis* task concerning a pattern appearing on the graphic.

$$
\left\{\begin{array}{cccc}
(\theta_8) & (\theta_9) & & (\theta_{15}) \\
\text{The climate} & \text{The suspension} & \ldots & \text{The audio} \\
\text{system had} & \text{had} & & \text{system} \\
\text{significantly} & \text{significantly} & & \text{significantly} \\
\text{more repairs in} & \text{more repairs in} & & \text{more repairs in} \\
\text{4 models,} & \text{3 models,} & & \text{1 model}
\end{array}\right\} \vdash
\begin{array}{c}
(\theta_{16}) \\
\text{The brand has a} \\
\text{high frequency} \\
\text{of repairs} \\
\text{outside the} \\
\text{power train}
\end{array}
$$

$$
\Uparrow \qquad\qquad \Uparrow \qquad\qquad \Uparrow \qquad\qquad \Uparrow
$$

$$
\left\{\begin{array}{cccc}
(\sigma_9) & (\sigma_{10}) & & (\sigma_{16}) \\
\text{A row labeled} & \text{A row labeled} & \ldots & \text{A row labeled} \\
\text{“Climate} & \text{“Suspension”} & & \text{“Audio system”} \\
\text{system” has 4} & \text{has 3 black} & & \text{has 1 black} \\
\text{black circles,} & \text{circles,} & & \text{circle}
\end{array}\right\} \vdash
\begin{array}{c}
(\sigma_{17}) \\
\text{The lower part} \\
\text{of the table is} \\
\text{blackish}
\end{array}
$$

FIGURE 101 An instance of consequence tracking supported by the system of frequency-of-repair tables. This underlies the transfer of a column-wise analysis of the perceptual pattern σ_{17}.

What guaranteed the validity of this task substitution? In our view, it was guaranteed by the consequence tracking relation in Figure 101.

Here, the task is to identify the set of more specific pieces of information that accounts for observed property θ_{16}. Thanks to the existence of σ_{17} that derivatively indicates θ_{16}, we can take the “lower route” in Figure 101: instead of directly analyzing trend θ_{16} to find $\theta_8, \theta_9, \ldots, \theta_{15}$, we can analyze perceptual pattern σ_{17} to find $\sigma_9, \sigma_{10}, \ldots, \sigma_{16}$ that make σ_{17} hold. Since σ_{17} and $\sigma_9, \sigma_{10}, \ldots, \sigma_{16}$ are perceptual properties, this part of the task is that of spatial analysis. Once $\sigma_9, \sigma_{10}, \ldots, \sigma_{16}$ are found, one can use the (derivative) meaning relation $\Rightarrow$ to read off information $\theta_8, \theta_9, \ldots, \theta_{15}$, and conclude that these are more specific facts that account for trend θ_{16}. Moreover, the validity of this conclusion is guaranteed by the existence of constraint $\{\theta_8, \theta_9, \ldots, \theta_{15}\} \vdash \theta_{16}$ in the upper part of Figure 101.

Note that the spatial analysis of perceptual pattern σ_{17} does not have to be unique. In addition to the row-wise analysis discussed above, we may conduct column-wise analysis, pointing out that the column labeled “’06” has four black circles and three mostly black circles in its lower part, that the column labeled “’07” has five black circles and one mostly black circle in its lower part, and so on. Or one may even conduct cell-wise analysis. Whatever the analysis may be, a valid transfer of spatial analysis to the target domain requires a target constraint $\Delta \vdash \delta_i$ to be tracked by a source constraint $\Gamma \vdash \gamma_i$, where δ_i and γ_i are typically more abstract types and Δ and Γ are sets of more specific types that make δ_i and γ_i hold.

Assuming that we are better at spatial analysis of directly perceivable patterns than at analysis dealing with non-perceptual information, we can see how the data analysis taking the lower route is more effective than that taking the upper route. This plausibly explains why people frequently resort to the indirect method when an appropriate graphic is supplied—people transfer a spatial analysis conducted on a diagram to obtain an analysis on the situation represented by the diagram.

This account implies that the transfer of spatial analysis is possible whenever there is such a lower route—namely, whenever a diagrammatic system supports the tracking of consequential constraint $\Delta \vdash \delta_i$ on the target domain by consequential constraint $\Gamma \vdash \gamma_i$ on the source domain. This implication has far-reaching consequences since there are many different cases in which diagrammatic systems have a consequence tracking properties. They include all the cases of free rides discussed in Chapter 2, meaning that whenever one can perform a free ride, one can also perform the transfer of spatial analysis. Furthermore, the abstraction tracking relation underlying the derived meaning typically implies multiple instances of consequence tracking and hence the possibility of spatial-analysis transfer. The present case involving σ_{17} and θ_{16} is one of these cases.

Admittedly, all these claims are only suggestive. Further justification of them, as well as the explanations of how they further clarify the cognitive potentials of diagrams, must be left for future research.

6.3 Aspect Shifting

Given a positive integer k, the *triangular number of* k is the sum of the first k positive integers $1, \ldots, k$. Thus, the triangular number of 1 is 1, that of 2 is $1 + 2 = 3$, and that of 3 is $1 + 2 + 3 = 6$. Let us write $T(k)$ for the triangular number of k. It is a theorem of natural number arithmetic that $T(k) = \frac{k \times (k+1)}{2}$ for every positive integer k.

The proof of this theorem is trivial in the case $k = 1$, since $T(k) = 1$ and $\frac{1 \times (1+1)}{2} = \frac{2}{2} = 1$. Now, suppose we want to prove this theorem for the case $k = 4$, i.e., to prove that $T(4) = \frac{4 \times (4+1)}{2}$. One "diagrammatic" proof goes as follows. We draw the dot triangle in Figure 102-a, which is composed of a column of one dot, a column of two dots, a column of three dots, and a column of four dots. Thus, the number of dots in this dot triangle is $1 + 2 + 3 + 4$, i.e., the triangular number $T(4)$. Then, we place a similar dot triangle on top of this triangle, so that the hypotenuses of the two triangles are aligned as they are in Figure 102-b. The newly added dot triangle is again composed of four columns of four dots, three dots, two dots, and one dot, so the number of dots in the upper triangle is $4 + 3 + 2 + 1$, i.e., the triangular number $T(4)$. Thus, the number of dots in the entire rectangle composed of the two triangles is

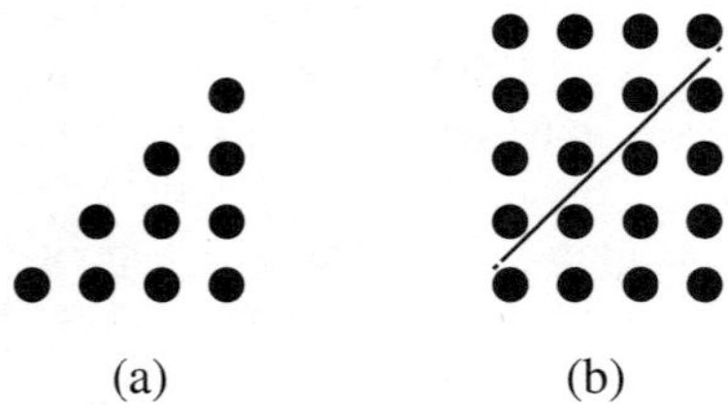

FIGURE 102 Dot diagrams used for a diagrammatic proof of $T(4) = \frac{4\times(4+1)}{2}$.

$T(4) + T(4)$, i.e., $2 \times T(4)$. Now, this dot rectangle can also be decomposed into four columns of dots, each containing five dots. Thus, the number of dots in this rectangle is $5 + 5 + 5 + 5$, i.e., $(4 + 1) + (4 + 1) + (4 + 1) + (4 + 1)$, i.e., $4 \times (4 + 1)$. Thus, the number of dots in the rectangle is both $2 \times T(4)$ and $4 \times (4 + 1)$, so $2 \times T(4) = 4 \times (4 + 1)$, and hence $T(4) = \frac{4\times(4+1)}{2}$.

It is clear, intuitively at least, that a proof of this form can be applied to any positive integer k greater than one. To state the common form of these proofs schematically, let us say that a dot triangle is *of magnitude* k if each of its sides consists of k dots. For example, the dot triangle in Figure 102-a is of magnitude four. Now the common form of the proof is:

(1) Draw two dot triangles of magnitude k vertically stacked with their hypothenuses aligned. Observe that the number of dots in the resulting dot rectangle is $2 \times T(k)$. Also observe that this dot rectangle can be decomposed into k columns of dots, each of which contains $k + 1$ dots. Thus, the number of dots in this rectangle is $k \times (k + 1)$. This makes $k \times (k + 1)$ equal to $2 \times T(k)$, and thus $T(k) = \frac{k\times(k+1)}{2}$.

Since it is obvious that a proof of this form applies to every $k > 1$, the diagrammatic proof of the general theorem often ends immediately after its instance for a particular k is shown, relying on our intuition that it can be applied to general cases. For the purpose of a formal proof, however, just pointing to this fact as being obvious is not sufficient. One needs to actually show that a proof of the form (1) applies to every k greater than one. That is not a trivial matter, partly because a dot diagram, as a physical figure, can only have a specific number of dots in it, and so any proof conducted on this basis is doomed to cover only one instance of the general theorem. This difficulty is sometimes cited as a fundamental weakness of this and similar diagrammatic proofs, and as support for the claim that diagrammatic proofs never attain the generality required for formal proofs of universally quantified propositions.

Resisting this argument, Giaquinto (2007) suggested a procedure free from this limitation, i.e., that of "abstracting certain features and ignoring some details" of a diagrammatic proof for a particular integer k. He claims that the "abstracted features might be sufficient as ingredients for a reliable procedure for reaching a true general proof" (p. 145).

In fact, Jamnik (2001) specifies an inference rule, called "the constructive ω-rule," that allows the proof of a universally quantified proposition under the condition that there is a uniform procedure, such as (1), that can be applied to each instance of the proposition. She calls such a procedure a *schematic proof*, and demonstrates that the system of dot diagrams supports schematic diagrammatic proofs for a large class of theorems in natural number arithmetic. She has even developed a program that could extract a schematic proof out of a couple of their instances and verify its general applicability.

Although we find the schematic-proof approach taken by Giaquinto and Jamnik quite promising in revealing the potential of diagrams for formal proofs of general mathematical propositions, this is not the issue that we focus in the following analysis. We are more interested in the very fact that there is a uniform procedure for diagrammatic proof that can be applied to each case of a certain general theorem, and the reason why there is such a procedure. Take the case of the general theorem, $T(k) = \frac{k\times(k+1)}{2}$. It seems that the uniform procedure described in (1) applies because we represent a positive integer in the form of dot shapes. The same uniform procedure would not apply if we represented an integer with a Kanji numeral, for example. Therefore, there must be something to this particular system of dot diagrams that supports the existence of a uniform procedure.

Our view is that a special kind of consequence tracking properties are responsible, but in order to explain this idea, we first clarify the semantics of the system of dot diagrams in accordance with Jamnik (2001). In this system, any dot shape denotes a positive integer, namely, the number of dots included in that shape. We write $[\![A_i]\!]$ to mean the integer that a dot shape A_i denotes. Then, the basic semantic convention of this system can be stated in the following way:

(2) If a dot shape A is composed of dots shapes $A_1, \ldots, A_n$, it indicates that integer $[\![A]\!]$ denoted by A is the sum of integers $[\![A_1]\!], \ldots, [\![A_n]\!]$ denoted by $A_1, \ldots, A_n$.

For example, the dot triangle in Figure 102-a is composed of a column of one dot, a column of two dots, a column of three dots, and a column of four dots, so it indicates that the integer denoted by this dot triangle is the sum of 1, 2, 3, and 4. Also, the fact that the dot rectangle in Figure 102-b is composed of four columns each of which contains five dots indicates that the integer denoted by this dot rectangle is the sum of 5, 5, 5, and 5.

Now, let k be an integer greater than one, and suppose we draw two dot triangles with magnitude k vertically stacked with their hypothenuses aligned. This necessarily results in a dot rectangle that is composed of k dot columns of magnitude $k + 1$. In other words, the constraint $\sigma_{18} \vdash \sigma_{19}$ holds for any integer $k > 1$, where:

(σ_{18}) Dot shape A is the vertical stack of two dot triangles of magnitude k with their hypothenuses aligned.

(σ_{19}) Dot shape A is a dot rectangular composed of k dot columns of magnitude $k + 1$.

Note that source types σ_{18} and σ_{19} respectively entail σ_{20} and σ_{21} below:

(σ_{20}) Dot shape A is composed of two dot shapes of size $T(k)$.

(σ_{21}) Dot shape A is composed of k dot shapes of size $k + 1$.

In other words, constraints $\sigma_{18} \vdash \sigma_{20}$ and $\sigma_{19} \vdash \sigma_{21}$ hold in the domain of dot diagrams. Furthermore, source type σ_{20} entails σ_{21}, making $\sigma_{20} \vdash \sigma_{21}$ another constraint in this domain.[29]

Due to the semantic conventions (2) explained above, source types σ_{20} and σ_{21} indicate target types θ_{17} and θ_{18}, respectively:

(θ_{17}) $[\![A]\!]$ is $T(k) + T(k)$.

(θ_{18}) $[\![A]\!]$ is $(k + 1)$ summed k times.

Note that target type θ_{17} entails target type θ_{18}, making $\theta_{17} \vdash \theta_{18}$ a constraint in the domain of integers.

Putting all this together, we can see the layered tracking of consequence in Figure 103 is supported by the present system of dot diagrams. Moreover, and more crucially, this layered tracking of consequence holds for every dot shape A and every integer k greater than one.

In our view, the consequence tracking properties of this general, layered kind account for the capacity of the system of dot diagrams to support the uniform proof procedure for the general theorem, $T(k) = \frac{k\times(k+1)}{2}$.

Let us see how this enables the schematic proof in (2) for every dot figure A and every integer k greater than one. Let k be an arbitrary integer greater than k. We assume that it has been already shown that dots triangles of magnitude k have $T(k)$ dots in them. The bottom constraint $\sigma_{18} \vdash \sigma_{19}$ guarantees that whenever we draw a dot figure A composed of two dot triangles of magnitude k that are vertically stacked, our figure A can also be decomposed into k columns of dots of magnitude $k + 1$. This guarantees the possibility of a viewer recognizing these two decomposition facts, σ_{18} and σ_{19}, about figure A. Since it has been proven that each dot triangle composing A has $T(k)$ dots in it, lefthand constraint $\sigma_{18} \vdash \sigma_{20}$ guarantees the possibility of recognizing

[29]Combined with still another constraint $\sigma_{19} \vdash \sigma_{18}$, this makes the constraints $\sigma_{18} \vdash \sigma_{21}$ and $\sigma_{19} \vdash \sigma_{20}$ also hold. These constraints, however, are not relevant to the inferences involved in the particular proof that we are concerned with.

(θ_{17}) $[\![A]\!]$ is $T(k) + T(k)$	$\vdash$	(θ_{18}) $[\![A]\!]$ is $(k + 1)$ summed k times
$\Uparrow$		$\Uparrow$
(σ_{20}) Dot shape A is composed of 2 dot shapes of size $T(k)$	$\vdash$	(σ_{21}) Dot shape A is composed of k dot shapes of size $k + 1$
$\perp$		$\perp$
(σ_{18}) Dot shape A is the vertical stack of two dot triangles of magnitude k with their hypothenuses aligned	$\vdash$	(σ_{19}) Dot shape A is a dot rectangle composed of k dot columns of magnitude $k + 1$

FIGURE 103 Layered consequence tracking relation underlying the aspect-shifting involved in the diagrammatic proof for the general theorem $T(k) = \frac{k\times(k+1)}{2}$. This relation holds for every dot shape A and every integer k greater than 1.

σ_{20}, which indicates fact θ_{17} due to the system's semantic conventions. Similarly, righthand constraint $\sigma_{19} \vdash \sigma_{21}$ guarantees the possibility of recognizing σ_{21}, which indicates fact θ_{18}. If the viewer successfully utilizes all these possibilities of recognition guaranteed by the consequence tracking in Figure 103 and reaches facts θ_{17} and θ_{18}, $T(k) = \frac{k\times(k+1)}{2}$ can be obtained through simple algebraic manipulations. Since the relation in Figure 103 holds for every integer k greater than one and for every dot figure A in this system, a proof of this form can be conducted for any integer greater than one on the basis of an arbitrary figure composed of two dot triangles that are stacked vertically. Thus, the set of consequence tracking properties of the form in Figure 103 underly the general applicability of the schematic proof (2).[30]

Giaquinto (2007) pointed out that the crucial discovery in the diagrammatic proof for the theorem "comes by viewing the final form in two ways, both as a rectangle of equal columns and as composed of the two equal triangles" (p. 157). He clearly explained that this process of viewing a figure in two ways plays a crucial role in many cases of diagrammatic proofs of mathematical theorems, from proofs involving dot diagrams to those involving

[30]Here, one does not have to be aware of the existence of this consequence tracking relation *per se* in order to follow the proof for a particular integer $k > 1$. It is sufficient for the viewer to recognize individual source types σ_{19}, σ_{20}, and σ_{21} when a diagram supporting σ_{18} is presented, and to understand that source types σ_{20} and σ_{21} indicate target types θ_{17} and θ_{18}, respectively. To understand the general applicability of this form of proof, however, the viewer probably need understand the decomposition constraint $\sigma_{18} \vdash \sigma_{19}$ and its generality over integers $k > 1$. This might prompt the *transfer* of the constraint $\sigma_{18} \vdash \sigma_{19}$ to the constraint $\theta_{17} \vdash \theta_{18}$ in the sense explained in Section 6.2, resulting in the understanding of the generality of the theorem.

simply arrays of symbolic expressions to those involving geometry diagrams. He called this general process *aspect-shifting*, and suggested that it is one of the most important methods of mathematical discovery.

In our analysis in Figure 103, the possibility of aspect-shifting is guaranteed by the existence of the lower-most constraint, $\sigma_{18} \vdash \sigma_{19}$, where source type σ_{18} stands for one way of decomposing figure A and source type σ_{19} stands for another. Because of this constraint, whenever one draws a diagram supporting decomposition type σ_{18}, it is possible to view the diagram as supporting decomposition type σ_{19}.

Our general hypothesis is then that aspect-shifting in a diagrammatic proof is based on consequence tracking properties, layered or un-layered, that involve constraints between two decomposition types of the same figure. In some cases such as the present example, those consequence tracking properties hold for a range of values of a certain parameter, supporting the application of a schematic proof to the corresponding range of cases of a universally quantified proposition. On this account, the ability of a diagrammatic system to support a proof of certain universally quantified propositions has its ultimate root in the universality of the decomposition constraints involved.

6.4 Law-Encoding Diagrams

Cheng (1996, 1997, 2002) has identified an important class of diagrams called *law-encoding diagrams*, which played crucial roles in scientific discoveries by Galileo, Newton, and other prominent scientists. A law-encoding diagram is characterized as "a representation that encodes the underlying relations of a law, or a system of simultaneous laws, in the structure of a diagram by the means of geometric, topological, and spatial constraints, such that the instantiation of a particular diagram represents a single instance of the phenomena or a particular case of the law(s)" (Cheng 1996: p. 146). This definition already suggests that systems of law-encoding diagrams have what we have characterized as consequence tracking properties.

To study the computational and cognitive properties of law-encoding diagrams, Cheng himself invented several systems of law-encoding diagrams, and Figure 104-a shows one of these diagrams, called a "one-dimensional property diagram" (Cheng 1996). This diagrammatic system is designed to express the relation of the velocities, the masses, and the directions of two erratic bodies colliding head on. In Figures 104-a and 104-b, for example, the lengths of arrows labeled "U_a" and "U_b" indicate the velocities (U_a and U_b) of bodies a and b before they collide, the ratio of the lengths of the lines labeled "m_a" and "m_b" indicates the ratio ($m_a : m_b$) of masses of a and b, and the lengths of arrows labeled "V_a" and "V_b" indicate the velocities (V_a and V_b) of the two bodies after collision. The directions of the arrows indicate the

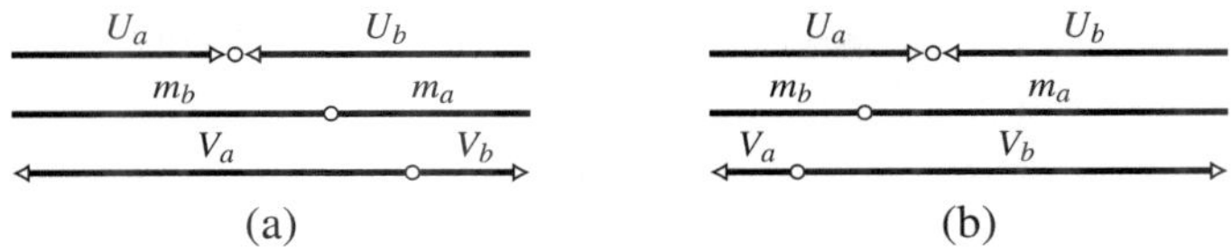

FIGURE 104 One-dimensional property diagrams, discussed by Cheng (1996) as typical law-encoding diagrams.

TABLE 2 Syntactic stipulations in the system of one-dimensional property diagrams, extracted from Cheng (1996).

S1	The tips of the arrows labeled "U_a" and "U_b" are adjacent, with a circle at the junction.
S2	The tails of the arrows labeled "V_a" and "V_b" are adjacent, with a circle at the junction.
S3	The mass lines are drawn end to end, with a circle at the junction.
S4	The left-to-right order of the mass lines is the reverse of the velocity arrows.
S5	The mass lines are drawn equidistant between two velocity lines, with their open ends vertically aligned with the tails of initial velocity arrows.
S6	The tails of the initial velocity arrows and the tips of the corresponding final velocity arrows must be in line vertically, so that the total distance between the tails of the arrows labeled "U_a" and "U_b" may be equal to the total distance between the tips of the arrows labeled "V_a" and "V_b."
S7	The three circles at the junction of the velocity lines and the mass lines must lie on a straight vertical or diagonal line.

directions of motions.

In addition to these semantic conventions, detailed syntactic stipulations are made on the forms of diagrams in this system, which we have summarized in Table 2.

From a logical perspective, this system of one-dimensional property diagrams supports numerous cases of consequence tracking, thanks both to spatial constraints on diagrams and to the set of syntactic stipulations that accompany them. Figure 105 shows just one of the many cases of consequence tracking supported by the system. The reader can check how this consequence tracking relation hods by conducting a "free-ride test": try to express premises θ_{19}, θ_{20}, and θ_{21} by closely following the syntactic stipulations in Table 2 and see if the diagram inevitably supports type σ_{25} and thus expresses θ_{22}.

Figure 104-b is the resulting diagram. Note that this diagram not only expresses θ_{22} (i.e., the velocity of a), but the the velocity of body b after it collides, as well as the directions of a's and b's motions after they collide. This means that three more consequence tracking properties hold in the system.

$$\left\{\begin{array}{ccc} (\theta_{19}) & (\theta_{20}) & (\theta_{21}) \\ U_a = 4, & U_b = 6, & m_a : m_b = 7:3 \end{array}\right\} \vdash \begin{array}{c} (\theta_{22}) \\ V_a = 2 \end{array}$$

$$\begin{array}{cccccc} \Uparrow & \Uparrow & \Uparrow & & & \Uparrow \end{array}$$

$$\left\{\begin{array}{ccc} (\sigma_{22}) & (\sigma_{23}) & (\sigma_{24}) \\ \text{The length of} & \text{The length of} & \text{The ratio of the} \\ \text{the arrow} & \text{the arrow} & \text{lengths of the} \\ \text{labeled ``}U_a\text{'' is} & \text{labeled ``}U_b\text{'' is} & \text{lines labeled} \\ 4, & 6, & \text{``}m_a\text{'' and ``}m_b\text{''} \\ & & \text{is 7:3} \end{array}\right\} \vdash \begin{array}{c} (\sigma_{25}) \\ \text{The length of} \\ \text{the arrow} \\ \text{labeled ``}V_a\text{'' is} \\ 2 \end{array}$$

FIGURE 105 An instance of consequence tracking supported by the system of one-dimensional property diagrams.

$$\left\{\begin{array}{ccc} (\theta_{22}) & (\theta_{23}) & (\theta_{21}) \\ V_a = 2, & V_b = 8, & m_a : m_b = 7:3 \end{array}\right\} \vdash \begin{array}{c} (\theta_{19}) \\ U_a = 4 \end{array}$$

$$\begin{array}{cccccc} \Uparrow & \Uparrow & \Uparrow & & & \Uparrow \end{array}$$

$$\left\{\begin{array}{ccc} (\sigma_{25}) & (\sigma_{26}) & (\sigma_{24}) \\ \text{The length of} & \text{The length of} & \text{The ratio of the} \\ \text{the arrow} & \text{the arrow} & \text{lengths of the} \\ \text{labeled ``}V_a\text{'' is} & \text{labeled ``}V_b\text{'' is} & \text{lines labeled} \\ 2, & 8, & \text{``}m_a\text{'' and ``}m_b\text{''} \\ & & \text{is 7:3} \end{array}\right\} \vdash \begin{array}{c} (\sigma_{22}) \\ \text{The length of} \\ \text{the arrow} \\ \text{labeled ``}U_a\text{'' is} \\ 4 \end{array}$$

FIGURE 106 Another instance of consequence tracking supported by the system of one-dimensional property diagrams.

Figure 106 shows another instance of consequence tracking in this system. It shows that this system not only supports the tracking of consequence from the initial velocities and the mass ratio to the final velocities, but also from the final velocities and the mass ratio to the initial velocities. In fact, the system supports the tracking of consequence from the initial and the final velocities to the mass ratio. Also, the system correctly tracks consequences for qualitatively different situations in collisions, as the diagrams in Figure 107 testify.

One obvious implication of these observations is that the system of one-dimensional property diagrams supports various free rides, letting the diagrams automatically express the value of a variable (e.g., the final velocity of a body) on the expression of values of other variables (e.g., the initial velocities and the mass ratio). This feature certainly contributes to the advantages of this system, but there is more to it than this. According to what

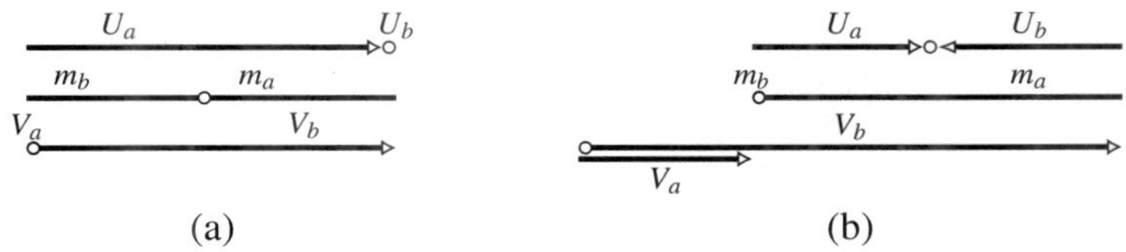

FIGURE 107 Other examples of one-dimensional property diagrams, representing the situations where (a) the initial velocity of body b is null, and (b) the mass of body b can be ignored compared to that of a (e.g., a pea and a planet).

we have learnt from the previous sections of this chapter, whenever a system has consequence tracking properties, it supports valid inferences based on envisaging, as well as valid inferences based on the transfer of spatial analysis. This suggests that the system of one-dimensional property diagrams has great potential for envisaging and spatial analysis transfers.

In fact, given the diagram in Figure 104-a, one can envisage a redrawing of the diagram that places the circle further to the left on the mass line than the present position so that the ratio of the lengths of the lines labeled "m_a" and "m_b" may become 7:3. If this envisaged redrawing involves no change in the lengths of the arrows on the initial velocity line, then the resulting diagram would look like Figure 104-b, and the circle on the final velocity line would be further to the left of the present position, making the length of the arrow labeled "V_a" much shorter. This change is mandated by the constraint in the lower part of Figure 105. By comparing the envisaged length of the arrow labeled "V_a" to its original length in Figure 104-a, one becomes aware of constraint $\{\sigma_{22}, \sigma_{23}, \sigma_{24}\} \vdash \sigma_{25}$ holding on the one-dimensional property diagrams. This kind of envisaging is supposedly cheap, and by repeating such envisaging "experiments," one may well familiarize oneself with these spatial constraints on one-dimensional property diagrams.

Now, thanks to the system's consequence tracking properties, these spatial constraints have their counterparts in the target domain, so that one can transfer each spatial constraint learnt in this way to learn a constraint governing the initial velocities, mass ratio, and final velocities of colliding bodies. Constraint $\{\theta_{19}, \theta_{20}, \theta_{21}\} \vdash \theta_{22}$ in the upper part of Figure 105 is an example of such a target constraint. In this way, the system of one-dimensional property diagrams can serve as a fertile platform for inference and learning.

Thus, from a logical point of view, an important clue that accounts for the efficacy of the system of one-dimensional property diagrams is its support of numerous instances of consequence tracking and its combined potentials for free rides, envisaging, and spatial analysis transfers. A preliminary survey has revealed that analysis along this line applies to other systems of law-encoding diagrams discussed by Cheng (1996, 1997, 2002) , including the systems of

Newton's motion parallelograms, his time-triangle diagrams, Galileo's free fall diagrams, his time-squared diagrams, his inclined-plane diagrams, Lewis' structure diagrams, and Cheng's own Amps-Volts-Ohms-Watts (AVOW) diagrams.

In fact, Cheng (1996) seems to have anticipated our analysis when he wrote:

> In contrast to other forms of diagrammatic representations, such as graphs, LEDs [law-encoding diagrams] exploit the variety and richness of geometric, topological, and spatial forms to express relations in the laws while using constraints so as to ensure that only lawful configurations of LEDs are permitted (Cheng 1996: p. 159) .

Here, he clearly had in mind what we have characterized as consequence tracking properties. Moreover, in his explanation of how law-encoding diagrams facilitate the discovery of laws, Cheng (1996) explicitly mentioned "the replacement of laborious sequences of logical or mathematical inferences by quick perceptual inferences that exploit the power of the visual system" (Cheng 1996: p. 158) . In our account, this "replacement" corresponds to the task substitutions achieved in the process of free rides, envisaging, and spatial analysis transfers.

As early as the 1990s, Cheng and Simon (1995) and Cheng (1996) went so far as to propose a computational system, called HUYGENS, of discovering a system of law-encoding diagrams suitable for spotting the regularities in the given data. We expect that the analysis of the logical basis of law-encoding diagrams sketched out in this section can be combined with their computational and psychological approaches to providing a fuller understanding of the rich cognitive potentials of law-encoding diagrams.

6.5 Chapter Takeaways

- In envisaging, we replace a physical drawing operation on a diagram with a mental drawing operation while obtaining the same free-ride effect of reading off a consequential piece of information.
- In aspect-shifting, we look at an element of a diagram in multiple ways to read off different pieces of information about the object denoted by the element.
- Spatial-analysis transfer lets us replace the task of analyzing abstract information with spatial analysis of the perceptual pattern in a diagram that indicates the abstract information in question.
- As a promising hypothesis, all of these can be considered to be special cases where consequence tracking properties of a diagrammatic system trigger characteristic cognitive processes.

7

Conclusion: Diagrams as the Subject of Scientific Studies

7.1 Commonality of the Four Logical Properties

The main thesis of this book has been that some of the most important cognitive potentials of diagrams are rooted in logically characterizable semantic properties of relevant systems of diagrams. Figure 108 gives an overview of these semantic properties, as they are characterized in Chapter 2–5 of this book. Each of the four columns shows the general form of a semantic property on which an important (positive or negative) cognitive potential of diagrammatic systems is based.

By putting together these properties in one place, we can see a striking commonality among them. That is, they all involve the projection of a a structural constraint in one form or another. Property (a), which underlies a free-ride potential, is the case where a consequential constraint $\Gamma \vdash \gamma_i$ on representations is projected; property (b), which underlies an over-specificity potential, is the case where a disjunctive constraint $\Gamma_1 \vdash \Gamma_2$ is projected; property (c), which underlies a consistency check potential, is the case where an inconsistency constraint $\Gamma \vdash \emptyset$ is projected; property (d), which underlies a meaning derivation potential, is the case where the cluster of structural constraints that makes up an abstraction relation $\gamma \bowtie \mathcal{G}$ are projected.

To make this observation a little more precise, let us introduce the general notion of "constraint projection" at this point:

Definition 10 (Projection of a Constraint) Let Γ_1, Γ_2 be sets of source types of a representation system $\mathcal{R}$. We say $\mathcal{R}$ *projects a constraint from* Γ_1 *to* Γ_2 if the following conditions hold:

1. $\Gamma_1 \vdash \Gamma_2$,

(a)	(b)	(c)	(d)
$\Delta \;\vdash\; \delta_i$	$\Delta_1 \qquad \Delta_2$	$\Delta \;\vdash\; \emptyset$	$\theta \;\bowtie\; \mathcal{D}$
$\Uparrow \qquad \Uparrow$	$\Uparrow \qquad \Uparrow$	$\Uparrow$	$\uparrow_{D} \qquad \uparrow_{1\text{-}1}$
$\Gamma \;\vdash\; \gamma_i$	$\Gamma_1 \;\vdash\; \Gamma_2$	$\Gamma \;\vdash\; \emptyset$	$\sigma \;\bowtie\; \mathcal{G}$
Tracking of consequences	Projection of disjunctive constraints	Tracking of inconsistencies	Tracking of abstraction relations
$\triangledown$	$\triangledown$	$\triangledown$	$\triangledown$
Free-ride potentials	Over-specificities	Consistency-check potentials	Meaning derivation potentials

FIGURE 108 Overview of the semantic properties underlying the positive and negative cognitive potentials of diagrams studied in this book.

2. there are sets Δ_1, Δ_2 of target types such that:
   (a) $\Gamma_1 \Rightarrow \Delta_1$,
   (b) $\Gamma_2 \Rightarrow \Delta_2$

It is not difficult to prove that each of the four semantic properties listed in Figure 108 is a special case of constraint projection. The cases of properties (a) and (b) are obvious. For the case of property (c), just observe that $\emptyset$ is collectively indicates $\emptyset$ in a trivial manner. So, (c) is the case where $\Gamma \vdash \emptyset$, $\Gamma \vdash \Delta$, and $\emptyset \Rightarrow \emptyset$. For the case of (d), recall that an abstraction relation $\sigma \bowtie \mathcal{G}$ has been defined in terms of the cluster of constraints holding between source type σ and the sets of source types in $\mathcal{G}$. So, it is clear that the tracking of $\theta \bowtie \mathcal{D}$ by $\sigma \bowtie \mathcal{G}$ entails the projection of the cluster of constraints that make up $\sigma \bowtie \mathcal{G}$.[31]

Intuitively, a system's projecting a constraint means that the expression of information in that system is *intervened* by that constraint. For suppose a representation system projects a constraint from Γ_1 to Γ_2. Then by definition, there are sets Δ_1, Δ_2 of target types such that $\Gamma_1 \Rightarrow \Delta_1$ and $\Gamma_2 \Rightarrow \Delta_2$. So, when we express the set of information Δ_1 by realizing the set Γ_1 of source types, we end up expressing some member of Δ_1. This is because, by definition again, the constraint $\Gamma_1 \vdash \Gamma_2$ holds and Γ_2 collectively indicates Δ_2. In case Γ_1 is empty, it means that Γ_1 is an inconsistent set, so that it is the case where the expression of Δ_1 by Γ_1 is prohibited entirely. Remember that $\Gamma_1 \vdash \Gamma_2$ is a

[31]To be more exact, the reader may want to verify that $\sigma \bowtie \mathcal{G}$ is equivalent to the holding of the constraint $\sigma \vdash \Sigma_c$ for every choice set Σ_c on $\mathcal{G}$. So, the tracking of an abstraction relation $\theta \bowtie \mathcal{D}$ by $\sigma \bowtie \mathcal{G}$ entails the projection of the constraints of the form $\sigma \vdash \Sigma_c$, where Σ_c is a choice set on the abstracted collection $\mathcal{D}$.

constraint involving source types of the system, i.e., a structural constraint on the arrangements of symbols in a representation. It typically arises from natural geometrical constraints interacting with artificial syntactic stipulations about symbols' arrangements. Projection allows this constraint to intervene the expressive work of the system in one way or another.

On this conception, what we have just seen with Figure 108 is that such an intervention can be both a good thing and a bad thing. Precisely because spatial constraints intervene our expression of information, our deductive inference for consequence or consistency can be facilitated; because of intervention, the information we read off from diagrams can be much richer than basic semantic conventions afford; because of intervention, we can make our expression over-specific and jeopardize the validity of our inference. The notion of constraint projection captures this core, the core semantic property that are responsible for important cognitive potentials of diagrams, both positive and negative.

This suggests a division of two broad classes of representation systems. One consists of those systems whose basic design allows such interventions so as to make the systems both enjoy their positive effects and suffer from their negative effects. The other consists of those systems whose basic design prevents constraint projections altogether so as to make the systems free from their positive and negative effects. Shimojima (2001) argues that this division corresponds to our intuitive terminological distinction between "graphic" and "linguistic" representation systems. I believe that this is broadly correct, but we no longer believe that it is such an important project to capture our terminological intuition on the basis of a formally defined concept ("conceptual analysis"). Rather than assessing whether the formal concept correctly captures the meaning of our age-old term "graphic," it seems more important to consider whether that concept, as a newly coined concept, can facilitate our understanding of the subject matter under investigation.[32] We will shortly see that the concept of constraint projection does an important job in this respect.

7.2 Implications of the Commonality

This brings us to the observation that constraint projection is a *fundamental* property of diagrammatic representation systems in general—fundamental in the sense that it underlies major cognitive potentials of diagrams, including negative ones.

One virtue of this observation is of course its ability to explicate seemingly diverse cognitive potentials in a unified manner. The potentials for free rides, over-specificity, consistency checks, and meaning derivations all have

[32] I owe this methodological view to the philosopher Ruth Millikan. See her *Varieties of Meaning* (Millikan 2004: p.36), for example.

their logical origins in the ways diagrammatic systems project structural constraints on their diagrams.

Another virtue of this observation is that it indicates a way of extending the present analysis to produce further explications. By investigating various ways diagrammatic systems project constraints, we may explicate, or even discover, other important cognitive potentials of diagrams than we have covered in this book. In the last chapter, we already indicated plausible ways of extending the analyses to explicate the semantic basis of other cognitive potentials, such as envisaging on diagrams, aspect-shifting, and transfers of spatial analyses. These extensions were all centered around the tracking of consequential constraints, but it is likely that other forms of constraint projection, listed or unlisted in Figure 108, will help us explicate further cognitive potentials of diagrams.

A final, perhaps farthest-reaching, virtue of our observation is its positive implications regarding the possibility of attaining a science of diagrams. We have identified a set of general semantic properties each of which is held by many diagrammatic representation systems. Moreover, these properties are special cases of an even more general property we called constraint projection. So, in this very general sense, most, perhaps all, systems that one may be inclined to consider "diagrammatic" are *similar*. Clearly, this similarity is not accidental, but supported by a ground in nature. There seems to be some common mechanism with which all members of the class are produced, and that should explain the similarity, namely, the existence of constraint projection properties in them.[33]

The search for this common mechanism is surely an important next step, but at this point, it is more important to realize the broader implications of our analysis. That is, the class of objects that diagram researchers have been studying are not just a conglomerate of unrelated things to which there is little substance. Rather, the class possesses the quality of a "natural kind" (Boyd 1991) that supports substantial inductions over its members. As such, diagrams make up a class of objects worth projecting a science over.

[33]Millikan classifies the reasons why a group of objects exhibit a natural, rather than accidental, similarity (Millikan 2000: chapter 2). The discussion in this and the following paragraph draws on her related account of what kinds of groups warrant useful scientific inductions.

Appendix A: Additional Analyses of Free Ride Potentials

Design Sketches

Recall from Subsection 2.1.5 that the story began with Quist drawing a sketch of land use, reproduced here in Figure 109-(a).

This sketch expresses the following pieces of information:

(θ_1) Three levels of ground are arranged in a step-like manner, with each upper level connected, at one end, to its lower neighboring level, at the nearest end, by a vertical elevation surface about 5 feet long.

(θ_2) The ground of each level is about 26 feet in depth.

These pieces of information are expressed by his sketch because the following states of affairs hold of it:

(σ_1) Three horizontal lines are arranged in a step-like manner, with each upper line connected, at one end, to its lower neighboring line, at the nearest end, by a vertical line about 0.5 inch long.

(σ_2) Each of the horizontal lines is about 2.6 inch in length.

Quist then modified the sketch to express the following information, producing the sketch in Figure 109-(b):

(θ_3) A rectangular building about 10 feet high and 19 feet deep stands on the second level, with the back side aligned with the elevation surface of the third level.

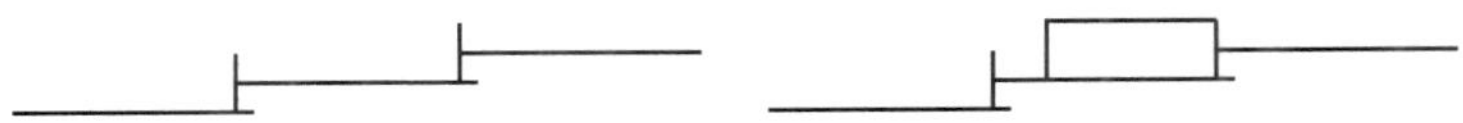

(a) (b)

FIGURE 109 Schematic reproduction of (a) Quist's first sketch and (b) Quist's modified sketch. Shown again.

This brought about the following state of affairs in addition to σ_1 and σ_2:

(σ_3) A rectangle about 1 inch high and 1.9 inch wide is on the second-level line, with one side aligned with the vertical line connecting the third-level line to the second-level line.

Due to a constraint governing the spatial configuration of the sketch, the following additional state of affairs is a consequence of states of affairs σ_1–σ_3:

(σ_4) The upper side of the rectangle is higher than the third-level line by about 0.5 inch.

But this state of affairs θ_4 indicates the following information:

(θ_4) The roof of the building rises about 5 feet above the ground of the third level.

Thus, the sketch ends up expressing information θ_4.

Figure 110 shows the mechanism of a free ride in this example. Information θ_4 was an unexpected consequence of the design ideas consisting of θ_1–θ_3, but it was conspicuously expressed in the sketch, and inspired further design ideas.

Block Drawings

WHISPER updates Figure 111-a with the following information:

(θ_5) The bottom surface of *B* inclines clockwise at 28°.

The resulting diagram (Figure 111-b) expresses the following information:

(θ_6) There is a gap between the lowest edge of *B* and the surface of *D*.

WHISPER thus obtains a free ride to θ_{13}.

What structural constraint on *Whisper*'s diagrams is responsible for this free ride? It is the constraint from source types:

(σ_5) The point labeled "*k*" is 1/4 inch leftward, and 3/16 inch upward from the top edge of the hexagonal icon labeled "*D*."

(σ_6) The bottom edge of the triangular icon labeled "*B*" is 5/8 inch long.

(σ_7) The top edge of the hexagonal icon labeled "*D*" is 7/8 inch long.

(σ_8) The point labeled "*k*" divides the bottom edge of the triangular icon labeled "*B*" in the ratio 6:11.

(σ_9) The bottom edge of the triangular icon labeled "*B*" inclines clockwise at 28°.

to source type:

(σ_{10}) There is a gap between the bottom edge of the triangular icon labeled "*B*" and the top edge of the hexagonal icon labeled "*D*"

$\left\{ \begin{array}{l} (\theta_1) \text{ Three levels of ground are arranged in a step-like manner, with each upper level connected, at one end, to its lower neighboring level, at the nearest end, by a vertical elevation surface about 5 feet long} \\ (\theta_2) \text{ The ground of each level is about 26 feet in depth} \\ (\theta_3) \text{ A rectangular building about 10 feet high and 19 feet deep stands on the second level, with the back side aligned with the elevation surface of the second level} \end{array} \right\} \vdash$ (θ_4) The roof of the building rises about 5 feet above the ground of the third level

⇑ ⇑

$\left\{ \begin{array}{l} (\sigma_1) \text{ Three horizontal lines are arranged in a step-like manner, with each upper line separated from its lower neighboring line by about 0.5 inch and connected to its near end with a vertical line} \\ (\sigma_2) \text{ Each horizontal line is about 2.6 inch in length} \\ (\sigma_3) \text{ A rectangle about 1 inch high and 1.9 inch wide is on the second-level line, with one side aligned with the vertical line connecting the third-level line to the second-level line} \end{array} \right\} \vdash$ (σ_4) The upper side of the rectangle is higher than the third-level line by about 0.5 inch

FIGURE 110 A consequence tracking property underlying the free ride based on the design sketches.

We ask the reader to check that this constraint hold on the WHISPER diagram. Figure 112 then specifies the consequence tracking property responsible for this case of free ride.

Observe that the inference provided by this free ride is *valid* but not *sound*: the information, θ_{12}, based on WHISPER's original calculation of block *B*'s initial incline is incorrect. The inference is nevertheless useful, since WHISPER modifies its original calculation on the basis of θ_{13}. Thus, WHISPER uses this free ride not to go from a piece of correct information to another piece of correct information, but to go from a piece of incorrect information to another piece of incorrect information that lets WHISPER realize the incorrectness of the first. The diagram serves as a test-bed for the first calculation. And it can serve as such because it obeys the bottom constraint governing its own structure.

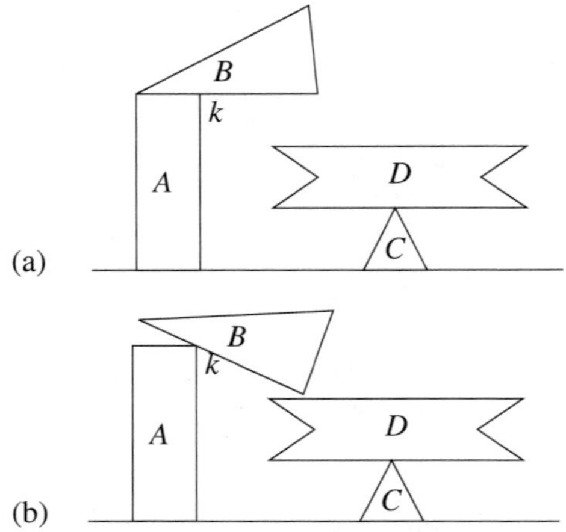

FIGURE 111 Block drawing (a) before expressing information θ_{12} about B's rotation, and (b) after the expression. Shown again.

This brings us to some very different instances of the free ride phenomenon than we have considered so far. Think of the cases of so-called "hardware simulation" where people use dummies for auto-accident tests, scale models of buildings for ventilation tests, and miniature river systems for hydrological tests. In such a simulation, the testers create a situation (the simulating situation) that stands for the real world situation (the simulated situation) whose behavior under certain circumstances we wish to know. The testers (1) adjust the simulating situation so that it represents the circumstances of the simulated situation that we are interested in, and then (2) observe what happens in the simulating situation. Steps (1) and (2) correspond to the steps of a free ride in which we (1′) operate on a representation so that it represents our assumptions, and (2′) observe the results produced under the structural constraints governing the representation. Thus, cases of accurate simulations correspond to cases of valid free rides, and cases of inaccurate simulations to cases of invalid free rides. From this perspective, then, the cases of free rides that we have seen are special instances of simulations that use diagrams as the simulating situations. Funt used the expression "experimenting on diagrams." More accurately, we were simulating with diagrams.

(θ_7) Point k is 1/4 feet leftward, and 3/16 feet upward from the top edge of hexagon D, (θ_8) The bottom edge of triangle B is 5/8 feet long, (θ_9) The top edge of D is 7/8 feet long, (θ_{10}) Point k divides the bottom edge of triangle B in the ratio 6:11, (θ_5) The bottom surface of B inclines clockwise at 28°	$\vdash$	(θ_6) There is a gap between the lowest edge of B and the surface of D
$\Uparrow$		$\Uparrow$
(σ_5) The point labeled "k" is 1/4 inch leftward, and 3/16 inch upward from the top edge of the hexagonal icon labeled "D," (σ_6) The bottom edge of the triangular icon labeled "B" is 5/8 inch long, (σ_7) The top edge of the hexagonal icon labeled "D" is 7/8 inch long, (σ_8) The point labeled "k" divides the bottom edge of the triangular icon labeled "B" in the ratio 6:11, (σ_9) The bottom edge of the triangular icon labeled "B" inclines clockwise at 28°	$\vdash$	(σ_{10}) There is a gap between the bottom edge of the triangular icon labeled "B" and the top edge of the hexagonal icon labeled "D"

FIGURE 112 A consequence tracking property underlying the free ride potential of block drawings.

Appendix B: Analysis of a Derivative Meaning with Round-Robin Tables

Recall from Subsection 5.1.7 that the system of round-robin tables supports the derivative meaning relation from σ_{11} to θ_{11} below.

(σ_{11}) More white circles appear in a row headed by "CHN" than in a row headed by "TUR."

(θ_{11}) The China team has won against more teams than the Turkey team has.

We take up this example since it helps us illustrate an interesting variation of abstraction tracking, namely, the *iterative* case of abstraction tracking.

Note that σ_{11} is an abstract condition, with several different ways of realizing it. One particular way is that five white circles appear in a row headed by "CHN" while four white circles appear in a row headed by "TUR." Using the abbreviations introduced in Subsection 5.2.1, we can model this "way" as the following set of conditions:

$$\Gamma_1 = \{\bigcirc_5(\text{"CHN"-row}), \bigcirc_4(\text{"TUR"-row})\}$$

Here is another way:

$$\Gamma_2 = \{\bigcirc_5(\text{"CHN"-row}), \bigcirc_3(\text{"TUR"-row})\}$$

If we keep enumerate all the particular ways in which σ_{11} holds in this manner, we obtain a collection consisting of 15 sets of source types including Γ_1 and Γ_2. Figure 113 shows this collection, which we call $\mathcal{G}$.

So we can say σ_{11} is an abstraction over $\mathcal{G}$:

(1) $\sigma_{11} \bowtie \mathcal{G}$

Just as σ_{11} is an abstract condition, θ_{11} is an abstract condition that has multiple ways in which it can be realized. Using the abbreviations introduced earlier, we can express two of these ways as the following sets:

$$\Delta_1 = \{W_5(\text{CHN}), W_4(\text{TUR})\}$$

$$\mathcal{G} = \left\{ \begin{array}{l} \{\bigcirc_1(\text{“CHN”-row}), \bigcirc_0(\text{“TUR”-row})\}, \{\bigcirc_2(\text{“CHN”-row}), \bigcirc_0(\text{“TUR”-row})\}, \\ \{\bigcirc_2(\text{“CHN”-row}), \bigcirc_1(\text{“TUR”-row})\}, \{\bigcirc_3(\text{“CHN”-row}), \bigcirc_0(\text{“TUR”-row})\}, \\ \{\bigcirc_3(\text{“CHN”-row}), \bigcirc_1(\text{“TUR”-row})\}, \{\bigcirc_3(\text{“CHN”-row}), \bigcirc_2(\text{“TUR”-row})\}, \\ \{\bigcirc_4(\text{“CHN”-row}), \bigcirc_0(\text{“TUR”-row})\}, \{\bigcirc_4(\text{“CHN”-row}), \bigcirc_1(\text{“TUR”-row})\}, \\ \{\bigcirc_4(\text{“CHN”-row}), \bigcirc_2(\text{“TUR”-row})\}, \{\bigcirc_4(\text{“CHN”-row}), \bigcirc_3(\text{“TUR”-row})\}, \\ \{\bigcirc_5(\text{“CHN”-row}), \bigcirc_0(\text{“TUR”-row})\}, \{\bigcirc_5(\text{“CHN”-row}), \bigcirc_1(\text{“TUR”-row})\}, \\ \{\bigcirc_5(\text{“CHN”-row}), \bigcirc_2(\text{“TUR”-row})\}, \{\bigcirc_5(\text{“CHN”-row}), \bigcirc_3(\text{“TUR”-row})\}, \\ \{\bigcirc_5(\text{“CHN”-row}), \bigcirc_4(\text{“TUR”-row})\}, \end{array} \right\}$$

FIGURE 113 The collection $\mathcal{G}$ of sets of source types, which exhausts all particular ways in which σ_{11} holds.

$$\mathcal{D} = \left\{ \begin{array}{l} \{W_1(\text{CHN}), W_0(\text{TUR})\}, \{W_2(\text{CHN}), W_0(\text{TUR})\}, \{W_2(\text{CHN}), W_1(\text{TUR})\}, \\ \{W_3(\text{CHN}), W_0(\text{TUR})\}, \{W_3(\text{CHN}), W_1(\text{TUR})\}, \{W_3(\text{CHN}), W_2(\text{TUR})\}, \\ \{W_4(\text{CHN}), W_0(\text{TUR})\}, \{W_4(\text{CHN}), W_1(\text{TUR})\}, \{W_4(\text{CHN}), W_2(\text{TUR})\}, \\ \{W_4(\text{CHN}), W_3(\text{TUR})\}, \{W_5(\text{CHN}), W_0(\text{TUR})\}, \{W_5(\text{CHN}), W_1(\text{TUR})\}, \\ \{W_5(\text{CHN}), W_2(\text{TUR})\}, \{W_5(\text{CHN}), W_3(\text{TUR})\}, \{W_5(\text{CHN}), W_4(\text{TUR})\}, \end{array} \right\}$$

FIGURE 114 The collection $\mathcal{D}$ of sets of conditions, which exhausts all particular ways in which θ_{11} holds.

$$\Delta_2 = \{W_5(\text{CHN}), W_3(\text{TUR})\}$$

Let $\mathcal{D}$ be the collection of all sets of conditions of this kind, exhausting all the ways in which θ_{11} holds. Then $\mathcal{D}$ has 15 members, just as $\mathcal{G}$ does (see Figure 114).

Thus, θ_{11} is an abstraction over $\mathcal{D}$ in the target domain. Therefore, we write:

(2) $\theta_{11} \bowtie \mathcal{D}$

We saw in Subsection 5.2.1 that the present system of round-robin tables supports a derivative meaning from source type $\bigcirc_n$(X-row) to target type $W_n(x)$. Note that sets Γ_1 and Δ_1 are in a one-to-one correspondence under this derivative meaning relation. So let us say Γ_1 *collectively indicates* Δ_1 *by the system's derivative meaning relation.* In fact, every member of $\mathcal{D}$ collectively indicates a unique member of $\mathcal{G}$ in this way, and every member of $\mathcal{D}$ is collectively indicated by a unique member of $\mathcal{G}$. In other words, $\mathcal{G}$ and $\mathcal{D}$ are in a one-to-one correspondence under the collective indication by the system's derivative meaning relation.

Thus, we see another case of abstraction tracking, as depicted in Figure 115. This time, however, collections $\mathcal{G}$ and $\mathcal{D}$ are not mediated by the system's basic semantic conventions, but by the meaning relation derived from them. If the abstraction tracking in Figure 87 is of the "basic" kind, the one in Figure 115 is of the "iterative " kind.

The difference between the basic and the iterative kinds, however, does not affect the validity of the informational relation that is derived from abstraction tracking. Since the informational relation from $\bigcirc_n(X\text{-row})$ to $W_n(x)$ is valid, and since every member of $\mathcal{G}$ is projected to some member of $\mathcal{D}$ by this informational relation, we find:

(3) For every set Γ in $\mathcal{G}$, if all members of Γ hold, then there is some set Δ in $\mathcal{D}$ whose members all hold.

From the two abstraction relations holding in the source and target domains, we obtain:

(4) If σ_{11} holds, then there is some set Γ in $\mathcal{G}$ whose members all hold.

(5) If there is some set Δ in $\mathcal{D}$ whose members all hold, then θ_{11} holds.

Combining (3), (4), and (5), we obtain:

(6) If σ_{11} holds, then θ_{11} holds.

A valid informational relation thus holds from σ_{11} to θ_{11}.

Even though a valid informational relation can be derived from an iterative case of abstraction tracking, that validity ultimately depends on basic cases of abstraction tracking. This is so because the meaning relation mediating an iterative case of abstraction tracking is that derived from basic cases. Since basic cases are mediated by the system's basic semantic conventions, these conventions are the origin of every instance of derivative meaning. In fact, we could also analyze the informational relation from σ_{11} to θ_{11} directly in terms of abstraction tracking mediated by the system's basic semantic conventions. In this case, we would have a collection of 386 sets of conditions as our $\mathcal{G}$ and a collection of the same size as our $\mathcal{D}$.

$$\theta_{11} \bowtie \left\{\begin{array}{l} \{W_1(\text{CHN}), W_0(\text{TUR})\}, \{W_2(\text{CHN}), W_0(\text{TUR})\}, \{W_2(\text{CHN}), W_1(\text{TUR})\}, \\ \{W_3(\text{CHN}), W_0(\text{TUR})\}, \{W_3(\text{CHN}), W_1(\text{TUR})\}, \{W_3(\text{CHN}), W_2(\text{TUR})\}, \\ \{W_4(\text{CHN}), W_0(\text{TUR})\}, \{W_4(\text{CHN}), W_1(\text{TUR})\}, \{W_4(\text{CHN}), W_2(\text{TUR})\}, \\ \{W_4(\text{CHN}), W_3(\text{TUR})\}, \{W_5(\text{CHN}), W_0(\text{TUR})\}, \{W_5(\text{CHN}), W_1(\text{TUR})\}, \\ \{W_5(\text{CHN}), W_2(\text{TUR})\}, \{W_5(\text{CHN}), W_3(\text{TUR})\}, \{W_5(\text{CHN}), W_4(\text{TUR})\}, \end{array}\right\}$$

↕ 1:1 correspondence under the collective indication by the derivative meaning relation

$$\sigma_{11} \bowtie \left\{\begin{array}{l} \{\bigcirc_1(\text{``CHN''-row}), \bigcirc_0(\text{``TUR''-row})\}, \{\bigcirc_2(\text{``CHN''-row}), \bigcirc_0(\text{``TUR''-row})\}, \{\bigcirc_2(\text{``CHN''-row}), \bigcirc_1(\text{``TUR''-row})\}, \\ \{\bigcirc_3(\text{``CHN''-row}), \bigcirc_0(\text{``TUR''-row})\}, \{\bigcirc_3(\text{``CHN''-row}), \bigcirc_1(\text{``TUR''-row})\}, \{\bigcirc_3(\text{``CHN''-row}), \bigcirc_2(\text{``TUR''-row})\}, \\ \{\bigcirc_4(\text{``CHN''-row}), \bigcirc_0(\text{``TUR''-row})\}, \{\bigcirc_4(\text{``CHN''-row}), \bigcirc_1(\text{``TUR''-row})\}, \{\bigcirc_4(\text{``CHN''-row}), \bigcirc_2(\text{``TUR''-row})\}, \\ \{\bigcirc_4(\text{``CHN''-row}), \bigcirc_3(\text{``TUR''-row})\}, \{\bigcirc_5(\text{``CHN''-row}), \bigcirc_0(\text{``TUR''-row})\}, \{\bigcirc_5(\text{``CHN''-row}), \bigcirc_1(\text{``TUR''-row})\}, \\ \{\bigcirc_5(\text{``CHN''-row}), \bigcirc_2(\text{``TUR''-row})\}, \{\bigcirc_5(\text{``CHN''-row}), \bigcirc_3(\text{``TUR''-row})\}, \{\bigcirc_5(\text{``CHN''-row}), \bigcirc_4(\text{``TUR''-row})\}, \end{array}\right\}$$

FIGURE 115 Abstraction tracking mediated by the derivative meaning relation for the system of round-robin tables. This relationship accounts for the valid informational relation from source type σ_{11} to target type θ_{11}.

References

Anscombe, Francis J. 1973. Graphs in statistical analysis. *American Statistician* 27:17–21.

Aristotle. 350 BC/1941. On memory and reminiscence. In R. McKion, ed., *The Basic Works of Aristotle*, pages 607–617. New York: Random House.

Aud, Susan, William Hussar, Frank Johnson, Grace Kena, Erin Roth, Eileen Manning, Xiaolei Wang, and Jijun Zhang. 2012. The condition of education 2012. Tech. Rep. NCES 2012-045, National Center for Education Statistics, US Department of Education, Washington, D.C.

Ballard, Dana H., Mary M. Hayhoe, Polly K. Pook, and Rajesh P. N. Rao. 1997. Deictic codes for the embodiment of cognition. *Behavioral and Brain Sciences* 20(4):723–767.

Barwise, Jon and John Etchemendy. 1987. *The liar: An essay on truth and circularity*. Oxford: Oxford University Press.

Barwise, Jon and John Etchemendy. 1991. Visual information and valid reasoning. In *Visualization in Teaching and Learning Mathematics*, pages 9–24. Washington, D.C.: Mathematical Association of America.

Barwise, Jon and John Etchemendy. 1994. *Hyperproof*. Stanford, CA: CSLI Publications.

Barwise, Jon and John Etchemendy. 1995a. Heterogeneous logic. In J. Glasgow, N. H. Narayanan, and B. Chanrasekaran, eds., *Diagrammatic Reasoning: Cognitive and Computational Perspectives*, pages 211–234. Cambridge, MA and Menlo Park, CA: The MIT Press and the AAAI Press.

Barwise, Jon and John Etchemendy. 1995b. Logic, proof, and reasoning. Unpublished.

Barwise, Jon and John Perry. 1983/1999. *Situations and Attitudes*. Stanford, CA: CSLI Publications. Originally published from MIT Press in 1983. Republished from CSLI Publications in 1999.

Barwise, Jon and Jerry Seligman. 1997. *Information Flow: the Logic of Distributed Systems*. Cambridge, UK: Cambridge University Press.

Bauer, Malcolm I. and P. N. Johnson-Laird. 1993. How diagrams can improve reasoning. *Psychological Science* 4(6):372–378.

Berkeley, George. 1710. *A Treatise Concerning Principles of Human Knowledge*. The Library of Liberal Arts. Indianapolis: The Bobbs-Merrill Company, Inc.

Bertin, Jacques. 1967/1973. *Semiology of Graphics: Diagrams, Networks, Maps*. Madison, WI: The University of Wisconsin Press. Originally published in 1967.

Bertin, Jacques. 1977/1981. *Graphics and Graphic Information*. Berlin: Walter de Gruyter. Originally published in 1977.

Boyd, Richard. 1991. Realism, anti-foundationalism and the enthusiasm for natural kinds. *Philosophical studies* 61(1):127–148.

Chandrasekaran, Balakrishnan, Bonny Banerjee, Unmesh Kurup, and Omkar Lele. 2011. Augmenting cognitive architectures to support diagrammatic imagination. *Topics in Cognitive Science* 3:760–777.

Cheng, Peter C.-H. 1996. Scientific discovery with law-encoding diagrams. *Creativity Research Journal* 9(2-3):145–162.

Cheng, Peter C.-H. 1997. Components of a cognitive theory of problem-solving and discovery with law encoding diagrams. In M. Anderson, ed., *Reasoning with Diagrammatic Representations: Papers from the 1997 AAAI Spring Symposium*, pages 85–93.

Cheng, Peter C.-H. 2002. Electifying diagrams for learning: principles for complex representational systems. *Cognitve Science* 26:685–736.

Cheng, Peter C.-H. and Herbert A. Simon. 1995. Scientific discovery and creative reasoning with diagrams. In S. M. Smith, T. B. Ward, and R. A. Finke, eds., *The creative cognition approach*, pages 205–228. MIT Press: Cambridge, MA.

Clark, Andy. 1997. *Being there: putting brain, body, and world together again*. Cambridge, MA: The MIT Press.

Clark, Andy. 2011. *Supersizing the mind: embodiment, action, and cognitive extension*. Oxford: Oxford University Press.

Clark, Andy and David Chalmers. 1998. The extended mind. *Analysis* 58(1):7–19.

Cleveland, W. S. 1994. *The Elements of Graphing Data*. Summit, NJ: Hobart Press.

Cummins, Robert. 1996. *Representations, Targets, and Attitudes*. Cambridge, MA: MIT Press.

Dennett, Daniel Clement. 1969. *Content and Consciousness*. London: Routledge and Kegan Paul.

Dwyer, Joseph R. and David M. Smith. 2012. Deadly rays from clouds. *Scientific American* 307(2):54–59.

Frege, Gottlob. 1879/1967. Begriffsschrift, a formula language, modeled upon that of arithmetic, for pure thought. In J. van Heijenoort, ed., *From Frege to Gödel: A Source Book in Mathematical Logic, 1879–1931*, pages 1–82. Cambridge, MA: Harvard University Press. Originally Published in 1879.

Funt, Brian V. 1980/1995. Problem-solving with diagrammatic representations. In J. I. Glasgow, N. H. Narayanan, and B. Chandrasekaran, eds., *Diagrammatic Reasoning: Cognitive and Computational Perspectives*, pages 33–68. Cambridge, MA and Menlo Park, CA: The MIT Press and the AAAI Press. Originally published in 1980.

Gattis, Merideth and Keith J. Holyoak. 1996. Mapping conceptual to spatial relations in visual reasoning. *Journal of Experimental Psychology: Learning, Memory, and Cognition* 22(1):231–239.

Gelernter, H. 1959. Realization of a geometry-theorem proving machine. In E. A. Feigenbaum and J. Feldman, eds., *Computers and Thought*, pages 134–152. New York: McGraw Hill.

Gelernter, H., J. R. Hansen, and D. W. Loveland. 1960. Empirical explorations of the geometry theorem machine. In E. A. Feigenbaum and J. Feldman, eds., *Computers and Thought*, pages 153–163. New York: McGraw Hill.

Giaquinto, Marcus. 2007. *Visual thinking in mathematics: an Epistemological Study*. Oxford: Oxford University Press.

Gilhooly, Kenneth J., Michael Wood, Paul R. Kinnear, and Caroline Green. 1988. Skill in map reading and memory for maps. *Quarterly Journal of Experimental Psychology* 40A:87–107.

Goldschmidt, Gabriela. 1994. On visual design thinking: the vis kids of architecture. *Design Studies* 15(2):158–174.

Gurr, Corin and Konstantinos Tourlas. 2000. Toward the principled design of software engineering diagrams. In *Proceedings of the 2000 International Conference on Software Engineering*, pages 509–518.

Gurr, Corin A. 1999. Effective diagrammatic communication: Syntactic, semantic and pragmatic issues. *Journal of Visual Languages and Computation* 10(4):317–342.

Gurr, Corin A., John Lee, and Keith Stenning. 1998. Theories of diagrammatic reasoning: Distinguishing component problems. *Minds and Machines* 8(4):533–557.

Guthrie, John T., Shelley Weber, and Nancy Kimmerly. 1993. Searching documents: Cognitive processes and deficits in understanding graphs, tables, and illustrations. *Contemporary Educational Psychology* 18:186–221.

Hammer, Eric. 1995. *Logic and Visual Information*. Stanford, CA: CSLI Publications and the European Association for Logic, Language and Information.

Hammer, Eric and Sun-Joo Shin. 1998. Euler's visual logic. *History and Philosophy of Logic* 19:1–29.

Hegarty, Mary. 1992/1995. Mental animation. In J. Glasgow, N. H. Narayanan, and B. Chanrasekaran, eds., *Diagrammatic Reasoning: Cognitive and Computational Perspectives*, pages 535–575. Cambridge, MA and Menlo Park, CA: The MIT Press and the AAAI Press. Originally published in 1992.

Hohauser, Sanford. 1982. *Architectural and Interior Models*. New York: Van Nostrand Reinhold Company, 2nd edn.

Hume, David. 1739. A treatise of human nature. In *The Philosophical Works of David Hume*, vol. 1. Boston: Little, Brown and Company.

Hutchins, Edwin. 1995. *Cognition in the wild*. Cambridge, MA: The MIT Press.

Jamnik, Mateja. 2001. *Mathematical Reasoning with Diagrams*. Stanford, CA: CSLI Publications.

Johnson-Laird, P.N. and B.G. Bara. 1984. Syllogistic inference. *Cognition* 16(1):1–61.

Kinnear, Paul R. and Michael Wood. 1987. Memory for topographic contour maps. *British Journal of Psychology* 78:395–402.

Kosslyn, Stephen M. 1994. *Elements of Graph Design*. New York: W. H. Freeman and Company.

Kulpa, Zenon. 2003. From picture processing to interval diagrams. Tech. rep., Instytut Podstawowych Problemów Techniki Polskiej Akademii Nauk, Warszawa.

Larkin, Jill H. and Herbert A. Simon. 1987/1995. Why a diagram is (sometimes) worth ten thousand words. In J. Glasgow, N. H. Narayanan, and B. Chanrasekaran, eds., *Diagrammatic Reasoning: Cognitive and Computational Perspectives*, pages 69–109. Cambridge, MA and Menlo Park, CA: The MIT Press and the AAAI Press. Originally published in 1987.

Lawson, Bryan. 1997. *How Designers Think: the Design Process Demystified*. Oxford: Achitectural Press.

Lemon, Oliver and Ian Pratt. 1997. Spatial logic and the complexity of diagrammatic reasoning. *Machine Graphics and Vision* 6(1):89–108.

Levesque, H. J. 1988. Logic and the complexity of reasoning. *Journal of Philosophical Logic* 17(4):355–389.

Lewis, David. 1975/1985. Languages and language. In *Philosophical Papers*, pages 93–115. Oxford: Oxford University Press. Originally published in 1975.

Lindsay, Robert K. 1988/1995. Images and inference. In J. I. Glasgow, N. H. Narayanan, and B. Chandrasekaran, eds., *Diagrammatic Reasoning: Cognitive and Computational Perspectives*, pages 111–135. Cambridge, MA and Menlo Park, CA: The MIT Press and the AAAI Press. Originally published in 1988.

Lohse, Gerald Lee. 1993. A cognitive model for understanding graphical perception. *Human-Computer Interaction* 8:353–388.

Lowe, Richard K. 1989. Speech strategies and inference in the exploration of scientific diagrams. *Educational Psychology* 9(1):27–44.

Lowe, Richard K. 1994. Diagram prediction and higher order structures in mental representation. *Research in Science Education* 24:208–216.

Luengo, I. 1995. *Diagrams in Geometry*. Ph.D. thesis, Department of Philosophy, Indiana University.

Marks, Joseph and Ehud Reiter. 1990. Avoiding unwanted conversational implicatures in text and graphics. In *Proceedings of the Eighth National Conference on Artificial Intelligence*, pages 450–455.

Miller, Nathaniel. 2007. *Euclid and his twentieth century rivals: Diagrams in the logic of Euclidean geometry*. Stanford CA: CSLI Publications.

Millikan, Ruth Garrett. 2000. *On clear and confused ideas: An essay about substance concepts*. Cambridge, UK: Cambridge UP.

Millikan, Ruth Garrett. 2004. *Varieties of Meaning: the 2002 Jean Nicod Lectures*. Cambridge, MA: The MIT Press.

Mineshima, K., M. Okada, and R. Takemura. 2012. A diagrammatic inference system with euler circles. *Journal of Logic, Language and Information* 21(3):365–391.

Mumma, John. 2010. Proofs, pictures, and Euclid. *Synthese* 175(2):255–287.

Narayanan, N. Hari, Masaki Suwa, and Hiroshi Motoda. 1995. Hypothesizing behaviors from device diagrams. In J. Glasgow, N. H. Narayanan, and B. Chanrasekaran, eds., *Diagrammatic Reasoning: Cognitive and Computational Perspectives*, pages 501–534. Cambridge, MA and Menlo Park, CA: The MIT Press and the AAAI Press.

Norman, Jesse. 2000. Differentiating diagrams: A new approach. In *Theory and Application of Diagrams: First International Conference, Diagrams 2000*, pages 105–116.

Novak, Gordon. 1988/1995. Diagrams for solving physical problems. In J. I. Glasgow, N. H. Narayanan, and B. Chandrasekaran, eds., *Diagrammatic Reasoning: Cognitive and Computational Perspectives*, pages 111–135. Cambridge, MA and Menlo Park, CA: The MIT Press and the AAAI Press. Originally Published in 1988.

Oberlander, Jon. 1996. Grice for graphics: Pragmatic implicature in network diagrams. *Information Design Journal* 8:163–179.

Olivier, Patrick. 2001. Diagrammatic reasoning: an artificial intelligence perspective. In A. F. Blackwell, ed., *Thinking with Diagrams*, pages 63–78. Dordrecht: Kluwer Academic Publishers.

O'Regan, Kevin J. 1992. Solving the "real" mysteries of visual perception: The world as an outside memory. *Canadian Journal of Psychology* 46(3):461–488.

Petre, Marian. 1995. Why looking isn't always seeing: readership skills and graphical programming. *Communications of the ACM* 38(6):33–44.

Petre, Marian and Thomas R. G. Green. 1992. Requirements of graphical notations for professional users: Electronics CAD systems as a case study. *Le Travail humain* 55:47–70.

Pinker, Steven. 1990. A theory of graph comprehension. In *Artificial Intelligence and the Future of Testing*, pages 73–126. Hilsdale, NJ: L. Erlbaum Associates.

Pylyshyn, Zenon W. 1973. What the mind's eye tells the mind's brain: A critique of mental imagery. *Psychological Bulletin* 80:1–24.

Pylyshyn, Zenon W. 1989. The role of location indexes in spatial perception: A sketch of the FINST spatial-index model. *Cognition* 32(1):65–97.

Pylyshyn, Zenon W. 2003. *Seeing and Visualizing: It's Not What You Think*. Cambridge, MA: The MIT Press.

Pylyshyn, Zenon W. and R. W. Storm. 1988. Tracking multiple independent targets: evidence for a parallel tracking mechanism. *Spatial Vision* 3(3):179–197.

Ratwani, Raj M., J. Gregory Trafton, and Deborah A. Boehm-Davis. 2008. Thinking graphically: Connecting vision and cognition during graph comprehension. *Journal of Experimental Psychology: Applied* 14(1):36–49.

Scaife, M. and Y. Rogers. 1996. External cognition: how do graphical representations work? *International Journal of Human-Computer Studies* 45(2):185–213.

Schön, Donald A. 1983. *The Reflective Practioner: How Professionals Think in Action*. New York: Basic Books.

Schwartz, Daniel L. 1995. Reasoning about the referent of a picture versus reasoning about the picture as the referent: an effect of visual realism. *Memory and Cognition* 23(6):709–722.

Shimojima, Atsushi. 2001. The graphic-linguistic distinction: Exploring alternatives. In A. F. Blackwell, ed., *Thinking with diagrams*, pages 5–27. Dordrecht, Netherlands: Springer.

Shimojima, Atsushi. 2002. A logical analysis of graphical consistency proofs. In L. Magnani, N. J. Nersessian, and C. Pizzi, eds., *Logical and Computational Aspects of Model-Based Reasoning*, pages 93–115. Berlin: Springer.

Shimojima, Atsushi and Dave Barker-Plummer. 2014. The Barwise-Seligman model of representation systems: A philosophical explication. Submitted to Diagrams 2014, The 8th International Conference on the Theory and Application of Diagrams.

Shimojima, Atsushi and Yasuhiro Katagiri. 2013. An eye-tracking study of exploitations of spatial constraints in diagrammatic reasoning. *Cognitive Science* 37(2):211–254.

Shin, Sun-Joo. 1991. A situation-theoretic account of valid reasoning with Venn diagram. In J. Barwise, J. M. Gawron, and S. Tutiya, eds., *Situation Theory and Its Applications*, pages 581–606. Stanford, CA: CSLI Publications.

Shin, Sun-Joo. 1994. *The logical status of diagrams*. Cambridge, UK: Cambridge University Press.

Sloman, A. 1971. Interactions between philosophy and artificial intelligence: the role of intuition and non-logical reasoning in intelligence. *Artificial Intelligence* 2(3–4):209–225.

Snow, John. 1855. *On the mode of communication of cholera*. John Churchill.

Spivey, Michael J. and Joy J. Geng. 2001. Oculomotor mechanisms activated by imagery and memory: eye movements to absent objects. *Psychological Research* 65(4):235–241.

Spivey, Michael J., Daniel C. Richardson, and Stanka A. Fitneva. 2004. Thinking outside the brain: Spatial indices to visual and linguistic information. In J. M. Henderson and F. Ferreira, eds., *The Interface of Language, Vision, and Action: Eye Movements and the Visual World*, pages 161–189. New York: Psychology Press.

Stenning, Keith and Robert Inder. 1995. Applying semantic concepts to analyzing media and modalities. In J. Glasgow, N. H. Narayanan, and B. Chandrasekaran, eds., *Diagrammatic Reasoning: Cognitive and Computational Perspectives*, pages 303–338. Cambridge, MA and Menlo Park, CA: The MIT Press and the AAAI Press.

Stenning, Keith and Jon Oberlander. 1995. A cognitive theory of graphical and linguistic reasoning: Logic and implementation. *Cognitive Science* 19(1):97–140.

Suwa, Masaki, John S. Gero, and Terry A. Purcell. 2000. Unexpected discoveries and S-Invention of design requirements: Important vehicles for a design process. *Design Studies* 21(4):539–567.

Tufte, Edward R. 1983. *The Visual Display of Quantitative Information*. Cheshire, CN: Graphics Press.

Tufte, Edward R. 1990. *Envisioning Information*. Cheshire, CN: Graphics Press.

Tufte, Edward R. 2003. *The Cognitive Style of PowerPoint: Pitching Out Corrupts Within*. Cheshire, CN: Graphics Press.

Wainer, Howard. 1992. Understanding graphs and tables. *Educaional Researcher* 21:14–23.

Index